Exploring Data Structures

The Analysis of Survey Data

Volume I

Exploring Data Structures

Edited by

COLM A. O'MUIRCHEARTAIGH

London School of Economics and Political Science

and

CLIVE PAYNE

Nuffield College, Oxford

JOHN WILEY & SONS

Chichester · New York · Brisbane · Toronto

Reprinted July 1978

Library of Congress Cataloging in Publication Data:

Main entry under title:

The Analysis of survey data.

Includes index.
CONTENTS: v. 1. Exploring data structures. –
v. 2. Model fitting.
1. Social surveys. 2. Multivariate analysis.
I. O'Muircheartaigh, Colm A. II. Payne, Clive.
HN29.A64 300'.1'8 76-951

ISBN 0 471 01706 X (Vol. 1)
ISBN 0 471 99426 X (Vol. 2)
ISBN 0 471 99466 9 (Set)

Typeset in IBM Journal by Preface Ltd, Salisbury, Wilts
Printed in Great Britain by The Pitman Press, Bath, Avon

To

Seosamh agus Éanna

Joseph and Marjorie

Preface

Three factors have transformed the analysis of survey data in the last decade: the development of new techniques for data analysis, the increased statistical sophistication of social scientists and the availability of powerful computers to research workers. We aim in this book to provide a practical and up-to-date guide to the processing and analysis of survey data for survey practitioners and students. The emphasis is on multivariate techniques, many of which were developed in other fields but which provide a means of extracting from the data answers to questions which social researchers have always asked but were prevented from answering due to the unavaliability of appropriate analytical techniques. In many cases descriptions of the techniques are available only in technical journals or advanced statistical texts. We attempt a coordinated and broadly non-theoretical treatment in the context of social surveys. The particular methodological and practical problems involved in applying the techniques to survey data are given special attention. Since computer programs are necessary to use many of the multivariate techniques, reference is made to the availability of those programs and to any special features or difficulties involved in their use.

This book is not comprehensive nor, we suggest, could it be. The scope of the subject is so broad that the best we could hope to achieve was a compact and usable guide. There is an extensive literature on survey design, sampling theory and statistical theory: three areas which have important implications for the analysis of survey data. Although we do not attempt to cover them here, they provide the framework within which the book is based, and where relevant, reference is made in the text to appropriate sources. The examples used are based on survey applications in a wide range of substantive areas.

The readers we had in mind in preparing this book are survey practitioners (academic, commercial and government) and students of social science and statistics. Survey practitioners today come from diverse backgrounds ranging from social philosophy to mathematics. This diversification is evident in the development of approaches to survey analysis. The influence of the mathematician can be seen in the proliferation of techniques derived from the work of theoretical statisticians. The effect of the social scientist has been to modify the mathematical techniques so that they can be used to answer substantively interesting questions. In this book we have tried to bring the theoretical and substantive approaches together by concentrating on applications of the methods to real problems of survey analysis.

It is still generally true that undergraduate courses do not include multivariate analysis. In statistics this neglect has been justified by the need to provide the student with a thorough grounding in statistical theory. However the increasing use of quantitative methods in the social sciences has led to a demand from the students in these subjects for a treatment of multivariate methods which does not require a detailed knowledge of statistical theory. This book is in part a response to this demand.

Complete consistency in the level of mathematical treatment has not always been possible, nor in some cases, we felt, desirable. We have tried to reconcile the need to present the topics in such a way that they are intelligible to the non-mathematical reader with the desirability of providing adequately for the more statistically sophisticated. In some chapters where we felt that they would be useful for some readers, complex derivations and other technical information have been presented in sections marked with an asterisk. Such sections can be omitted without loss of continuity.

Ideally the methods of analysis should be an integral part of the survey design. In defining the survey objectives the researcher should consider what analytical techniques are available and should take into account in planning the data collection the type of data and level of measurement required for the techniques chosen. Frequently however, the data are collected prior to the detailed planning of the analysis and in such cases the choice of technique is constrained. The structure of the book permits its use both at the design stage and during the analysis.

Two factors influence the way in which social scientists approach problems of statistical analysis. First is the degree to which the level of knowledge in their substantive area permits the formal specification of a model prior to analysis. The second is the extent to which they are familiar with classical statistical estimation and inference procedures. The two volumes of this book reflect the different approaches which thus arise in survey analysis. On the one hand those who are dealing with relatively uncharted substantive areas will find that the techniques presented in Volume 1 are more immediately relevant as these deal with the exploration and identification of structures in the data. On the other hand the classical statistician or the social scientist who is concerned with hypothesis testing will find that the model fitting and testing procedures described in Volume 2 are closer to his needs. Those dealing with problems of exploration however may find that the structures identified can be used to fit more formal models using the techniques of Volume 2. Conversely, the model fitter may find that the abundance of data may constrain his analysis and force him to reduce the dimensionality of his problem. Many of the techniques in Volume 1 provide such a means of data reduction and summary.

Although each volume can be used independently there is a unifying structure to the book. Chapters 1 and 2 of the first volume provide the context within which the work is based. Chapter 1 has a special role in placing statistical analysis in the context of survey research with particular reference to the analytical methods presented while Chapter 2 deals with practical considerations of the computer processing of survey data. The rest of Volume 1 deals with a battery of multivariate techniques concerned with exploring data structure in an inductive way. Here the prior contribution of the substantive theory is in determining which variables are to be observed but not in specifying the precise functional form of the relationships between them.

Volume 2 begins with a chapter on statistical estimation and hypothesis testing which provides the foundation on which the substantive chapters are based. Chapters 2 to 5 deal with model fitting. The specification of models implies far more prior knowledge than exists in the situations discussed in Volume 1. The substantive theory must be considerably more exact in order to specify not merely the variables of interest but also the relationships between them. The main advantage of specifying the model *a priori* is that it is possible to test the fit of the model to the data. The main disadvantage is that specifying and fitting an inappropriate model may mask or obscure important relationships of other kinds in the data. The last two chapters complete the context of statistical analysis of survey data begun in Chapter 1 of Volume 1 by considering two underlying features of survey analysis. Chapter 6 deals with the effect of complex sample design on the correlation coefficient on which many multivariate techniques are based. The final chapter describes the sources, identification and effect of response errors and considers the substantial impact of these errors on inferences from survey data.

Our overall approach is an inductive one. 'Every solution of a problem raises new unsolved problems; the more so the deeper the original problem and the bolder its solution' (Popper, 1963). But the futility of seeking a complete solution has long been recognized. Perhaps we should take as our text the words of Xenophanes:

> 'The Gods did not reveal, from the beginning,
> All things to us; but in the course of time,
> Through seeking we may learn, and know things better.'

C.A. O'M.
C.D. P.

Contributors

A. P. M. COXON — *Department of Sociology, University College of Wales, Cardiff*

B. S. EVERITT — *Institute of Psychiatry, London*

A. FIELDING — *Department of Statistics, London School of Economics and Political Science*

C. L. JONES — *Department of Sociology, University of Edinburgh*

A. C. MCKENNELL — *Faculty of Social Studies, University of Southampton*

C. A. O'MUIRCHEARTAIGH — *Department of Statistics, London School of Economics and Political Science*

C. PAYNE — *Nuffield College, Oxford*

C. C. TAYLOR — *Institute of Psychiatry, London*

Contents: Volume I

Outline of Contents: Volume II
Model Fitting

Chapter 1

Statistical Analysis in the Context of Survey Research

C. A. O'Muircheartaigh

1.1 STAGES OF A SURVEY

The eventual output of a survey is a set of quantitative results, or a set of statements based on such results. These are generated by a succession of stages and decisions each of which has implications for and an effect on the meaning and usefulness of the output. A social survey differs in many respects from other types of research and it is these special features of surveys which are our main concern here. The stages of a survey are interrelated and should not in practice be considered independently of one another. However their main features can be described in sequence.

The extent and content of the *survey population* must first be defined. In this book we restrict ourselves to studies of human populations. The elements and the population must be defined together since the population is simply the aggregate of the elements. The definition must cover content, location and time. For instance the population could be defined as all males between the ages of 18 and 64 resident in the United Kingdom in 1976. There may be a difference between the target population and the survey population which is actually reached. It may not be possible, for example, to cover those elements of the population living in large institutions or those without a permanent residence.

The *nature of the survey variables* to be observed also affects all stages of the survey. Many different types of variables will be included but for each of them the level of measurement and the unit of measurement must be defined. Some of the variables, such as age, sex and marital status may be easily defined. In the case of other objective variables, such as number of rooms in the house and standard of lighting, careful definition of categories may be necessary. For attitudinal variables the problems will be more severe, that is, whether to depend on a single item or whether to construct an attitude scale from a set of items and what level of measurement to aim for, and in many cases there will be no perfect solution.

The *methods of data collection* transform the survey variables into observations and the methods which are used determine the nature of the data. Many methods are currently in use, ranging from participant observation at one extreme, through personal interviews, self-completion and mail questionnaires, to documentary sources at the other. The survey population and the survey variables together will affect the choice of data collection method.

Although it is possible in principle to observe all the elements in the population, social surveys usually deal with a sample from the population. The procedure for selecting the elements to be included in the sample is called the *selection process*. There are two fundamentally different kinds of selection process: *probability sampling*, in which each element in the population has a known non-zero probability of selection, and *non-probability sampling*, which includes all other procedures. The classical techniques of statistical inference are relevant only to probability sampling.

The *sample design* is the set of rules which defines the selection process. The simplest is simple random sampling where each combination of elements in the population has the same probability of being included in the sample. Departures from simple random sampling affect the inferences which can be made from the sample statistics to the population parameters

The *estimation process* is the set of rules which enables us to compute the values of the sample statistics. The precision of these estimates is determined by the population structure and the selection process. The effects of the sample design will vary for different types of estimators, and for different analytical techniques.

The *methods of analysis* provide a means of transforming the raw survey data (the observations) into a meaningful set of results. The methods therefore are determined jointly by the nature of the survey variables and the survey objectives. The objectives may range from a simple description of the population to an examination of a complex set of relationships between the population elements and the set of survey variables.

The purpose of this book is to present a description and evaluation of statistical methods of data analysis. In this chapter the other aspects of the total survey design are considered briefly, particularly in relation to their effect on the choice and utilization of particular analytical techniques. The survey population, the choice of survey variables and the methods of data collection influence the selection and estimation processes and all affect the methods of analysis. In an ideal situation the choice of methods of analysis should be made as an intrinsic part of the survey design. In most surveys this is not the case and the survey analyst is faced with the task of deciding which techniques are appropriate and available for a particular set of data.

1.2 THE NATURE OF THE SURVEY VARIABLES

There is a considerable difference between sociological theory and empirical social research. To the theorist, events can be repeated and properties can be assumed to be constant. In real life, no event can ever be exactly repeated nor is any measurable property completely constant. The degree to which the assumptions of the theorists can be met will vary from one case to another. In theory, we have repeatable events with constant properties. In practice, we have the survey conditions, observations and predictions. The variables normally used in social surveys differ in the degree to which they conform to a theoretical framework; they range from objective characteristics at one extreme to attitudes and emotions at the other. Almost all the properties or constructs in the theory will be represented by indicators in practice. For example, instead of measuring authoritarianism, we record the responses of the individuals to a set of attitudinal items dealing with authoritarianism. There is an inherent gap between the theoretical constructs and the operational definitions. The operationalism controversies are adequately discussed elsewhere (see, for example, Blalock and Blalock (1968)). It is important however to distinguish between different types of empirical measurement to appreciate their limitations.

1.2.1 Levels of Measurement

A *scale* may be defined (Blalock and Blalock, Ch. 3) as consisting of three components:

(1) an empirical event (such as a person or an inanimate object);
(2) a number;
(3) the rule or set of rules that links the event and the number (the mapping rules).

The investigator specifies a property in terms of which a set of observable events might be specified, and defines rules which enable him to distinguish between different statuses of the property. The meaning of the numerical values allocated to the different statuses is unaltered by changes, the limits of which can be specified in each case. This is the concept of an *admissible transformation*, i.e. any transformation of the numerical values assigned in the scale which does not affect the meaning of the scale values. Levels of measurement (types of scale) can be distinguished by the set of admissible transformations for that scale.

The lowest level of measurement is the *nominal scale*. This is simply the substitution of names or symbols for real objects. The objects are allocated to categories which must satisfy the following conditions (Coombs, 1953):

(i) Each object or individual either belongs to a category or it does not [*equivalence* or *non-equivalence*].
(ii) The equivalence relationship must be symmetric and transitive.

These are necessary conditions for all higher levels of measurement. The numerical assignments for nominal scale variables are arbitrary and can be subjected to any transformation which preserves the equivalence relationship. Thus the methods of analysis appropriate for such data must take this into account. The use of frequency counts, calculation of the mode, the use of chi-square or coefficients of contingency are permissible, but no analytic technique should be used whose results would vary according to the assignment of numbers to the classes. The numbers merely name the classes.

An *ordinal scale* implies that the numerical values assigned to the observations indicate that the observations have more or less of the underlying property.

Socio-economic status might be an example of such a scale. The assignment of numerical values to the classes simply reflects the ordering. With the nominal scale the only restriction on the numbers is that the observations receiving the same number are alike in terms of the property. This also applies to the ordinal scale with the additional requirement that the numbers assigned must correspond to the order of the classes on the scale. Thus, all monotonic transformations are admissible for ordinal scales. The methods of analysis which are appropriate are those which use only the ordinal properites of the numbers. Non-parametric and ranking methods in general are permissible. Hollander and Wolfe (1973) present a good discussion of standard non-parametric statistical methods.

When dealing with more than one ordinal scale, the combination of such scales may produce a *partially ordered* scale. If for example, socio-economic status were determined by a combination of income and years of education, the rules might be as follows:

(1) $A > B$ if both education $(A) >$ education (B)
and income $(A) >$ income (B)

(2) $A = B$ if both education $(A) =$ education (B)
and income $(A) =$ income (B)

(3) $A \neq B$ if education $(A) >$ education (B)
income $(A) <$ income (B)

In case (3) no comparison on the scale is possible without some formal equation between income and education in terms of the scale.

Interval scales are characterized by the fact that in addition to the properties of the ordinal scale the numerical values contain information on the size of the intervals between positions on the scale. This requires a distance function which assigns a real number to all pairs of numbers in the ordered set. The main restriction which must be added to those imposed for the nominal and ordinal scales is that the difference in the numbers assigned to two positions on the scale must reflect the *amount* of difference in the underlying property. The admissible transformations for the scale are linear transformations of the form $y = ax + b$, where a is a positive number and b can be positive or negative. The appropriate methods of analysis therefore are much more powerful. With interval scales it is possible to define a unit of measurement and to describe distances on the scale in terms of that unit. Thus, the calculation of means, variances, product-moment correlations and the application of parametric statistical techniques are all appropriate.

The highest level of measurement is the *ratio scale*. The essential difference between interval and ratio scales is that for the latter an absolute zero can be defined, a value which indicates the absence of the property. Thus the numerical values assigned to scale positions reflect not only identity, order and differences in magnitude but also absolute magnitude. The only transformations admissible are those of the form $y = ax$ where a is a positive number. All types of statistical techniques are appropriate for use with such scales.

1.2.2 Relevance to Survey Variables

The variables observed in many social surveys will cover the whole range from nominal through to ratio scales. Demographic and physical measurements will frequently satisfy the constraints of the ratio scale. Age and income are examples of such variables. However the majority of the variables will be measured at a lower level. Attitude measurement in most cases will not be above the ordinal level, although in the analysis we may wish to assume that the observations can be treated as interval level scales. The methods of analysis used are crucial in this respect. What the researcher 'finds out' from the data is a function not only of the information in the data but also of the process by which it is extracted. It is important to distinguish between the postulates and assumptions underlying the analysis and the conclusions drawn from the results. For example, if the instruction to the respondent is of the form 'Place the following k items in rank order', then the data collected cannot allow other interpretations to emerge. Similarly in the analysis the data can be forced to conform to an ordinal scale simply by classifying as error that part of the data which does not conform. Thus no property of the data can be said to hold unless the methods of collecting and analysing the data permit the alternatives to manifest themselves (Coombs, Dawes and Tversky, 1970). It is generally more fruitful to use a scale at the highest level of measurement satisfied by the data. In attitude measurement the minimum requirement is usually an ordinal scale. Chapter 7 presents a discussion of attitude scale construction in this context.

Many of the variables dealt with in surveys are measured on a nominal scale. Sex, marital status, race and political party are examples. It is important therefore to have analytic methods available which can deal with combinations of variables which are measured at different levels without violating the characteristics of each scale. Many such techniques have been developed and some are discussed in detail in Chapters 5, 6, 7 and 8 and in Chapter 4 of Volume 2.

One further characteristic of the measurement process in surveys is important. Frequently the variables observed are simply approximations to the variables of interest: the theoretical constructs or dimensions which form the basis of the sociological or psychological theory. The manifest space, that is, the observed or measured variables, serves as an indicator or guide to the latent space. Many of the statistical techniques described in this book represent attempts to deduce the latent dimensions from the observations. Chapters 3 through 6 present approaches to this problem.

1.3 THE METHODS OF DATA COLLECTION

Most survey design textbooks give a formal discussion of the methods of data collection which are commonly used. Moser and Kalton (1971) provide an excellent treatment. In this section the discussion is confined to the implications of problems arising in mail and interview surveys for the analysis of the data.

The first distinction to be drawn is between errors of observation and those due to non-observation. The latter arise due to noncoverage – for example, incomplete frames – and nonresponse. Errors of observation occur through inaccurate obtaining and recording of observations. The second distinction is between constant and variable errors. Constant errors or *biases* account for the difference between the population value and the true value. 'This difference delineates the gap between the defined survey objectives and the entire set of essential survey conditions' (Kish, 1965). Variable errors may be represented in terms of sums of squares of deviations from the true values.

1.3.1. Noncoverage

Failure to include some parts of the defined survey population in the frame constitutes noncoverage. By virtue of this exclusion some elements in the population have a zero probability of inclusion in the survey. Thus that part of the population is completely neglected in the survey results. There is a difference between noncoverage and situations where some sections of a population are explicitly excluded from consideration on other grounds. For example, a decision may be taken to confine a study of adults to those living in households. Those in institutions may be excluded on practical grounds. In such cases, the survey population is *difined* as adults living in households, and it is to this survey population that the results refer. Noncoverage consists of failure to include elements which the survey objectives wish to include. In most cases noncoverage arises through faults in the sampling frame but inadequate field procedure can result in a similar neglect of population elements.

A discussion of noncoverage is more appropriate to texts dealing with survey design, but even after the completion of a survey some information on the extent of noncoverage may be obtainable from sources outside the survey itself. Census information may provide comparable figures for some demograghic aspects of the survey coverage. It is important that any such figures should refer to the same units and to the same time period. The effect of noncoverage is very difficult to estimate but as in the case of nonresponse discussed below, the effect may well be considerably less for means than for simple expansion estimates of totals.

1.3.2. Nonresponse

In sample surveys where the respondents are under no obligation to cooperate, it will normally be the case that data will not be collected for all the designated units. This is particularly true if only one attempt is made to collect information from each unit. Data will be missing for some part of the population and the problem arises as to what conclusions are legitimate from such surveys or alternatively what means are available to gain insight into the part of the population not covered. The problem of nonresponse is common both to complete enumerations and sample surveys. In order to appreciate the significance of nonresponse it is necessary to classify it according to its source.

Sources of nonresponse

No single classification of nonresponse is suitable for all surveys. The aim should be to link together the types of nonresponse which are similar in effect and to distinguish between those whose effects on the results may differ. The classification below is based on Kish (1965). Moser and Kalton (1971) present an alternative.

(1) Not-at-homes (NAH's): the amount of nonresponse due to this cause will vary with the nature and habits of the respondents (e.g. farmers are more easily contacted during the day than industrial workers), with the timing of the calls, and with the location of the interview. The timing is relevant only in the case of personal interviews and does not apply to mail surveys. For interview surveys a distinction may be made between 'nobody at home' and 'respondent absent' situations. Kish suggests the term *temporarily unavailable* for this group.

(2) Refusals: these are due usually to the disposition of the respondent and will vary with the approach used, the sponsorship of the survey and the subject matter. The term *unobtainable* may be applied to this type of nonresponse, since repeating the approach will not change the situation.

(3) Incapacity to respond: this may be due to illness, illiteracy, or language difficulties and will typically not be a large group.

(4) Not found: this category includes vacant houses, incorrect addresses, and may include respondents who have moved from their listed address and interviews which have not been attempted for some reason.

(5) Lost schedules: refers to information lost or destroyed after it had been collected. It should not constitute a large proportion of the nonresponse.

The classification above refers to a complete absence of information. However in many cases, there will be a failure to record particular items of information in otherwise completed schedules. The sources of these omissions may be similar to those in the categories above. The methods of dealing with these are discussed in Chapter 2.

An Example

In each survey, particular sources of non-response will be dominant. These will vary from one survey to another. This example presents the classification of responses for a national study of occupational mobility carried out in Ireland in 1973 by Jackson, Iutaka *et al.* The survey population was defined as all males aged between 18 and 64 living either in Northern Ireland or the Republic of Ireland in 1973. The contact points for the interviews were 'male' names selected from the Register of Electors. The sample design is reported in O'Muircheartaigh and Wiggins (1976) from which Table 1.1 is adapted.

An examination of the categories shows that they fall into two conceptually different classes. The first class consists of categories 1 through 7 and includes all those contacts which did not generate eligible respondents. Category 1 is a special case in which those living in institutions were defined as not belonging to the survey

Table 1.1. Response categories in an occupational mobility study (Reproduced with permission from C. A. O'Muircheartaigh and R. D. Wiggins, *Economic and Social Review* (1976))

	All contacts		Eligible respondents	
Category	Frequency	Relative frequency	Frequency	Relative frequency
1. Institution	150	1.9	–	–
2. House demolished	37	0.5	–	–
3. House vacant	176	2.2	–	–
4. Person female	34	0.4	–	–
5. Person dead	150	1.9	–	–
6. Person moved	601	7.4	–	–
7. Person wrong age	1109	13.7	–	–
8. Not at home	470	5.8	470	8.0
9. Refusal of all information	136	1.7	136	2.3
10. Refusal of interview	552	6.8	552	9.5
11. Not contacted	10	0.1	10	0.1
12. Interview completed	4673	57.8	4673	80.0
Total	8098	100.0	5841	100.0

population. The movers in category 6 were covered by another aspect of the sample design. The response rate should be calculated excluding these categories and should deal with responses and nonresponses among the eligibles only, i.e. categories 8 through 12.

The categories in the second class do not all have the same significance. The effect of nonresponse on the survey estimates is caused by differences between the respondents and the nonrespondents. Refusals are more likely to differ from the respondents than are the not-at-homes for some variables, and the effect of nonresponse will vary according to the parameter of the population being estimated.

Effect of nonresponse

Consider the population to be divided into two parts, respondents and nonrespondents. Let N denote the population size and N_1 and N_2 the number of respondents and nonrespondents respectively. Let $W_1 = N_1/N$ and $W_2 = N_2/N$ denote the proportions, and $\bar{Y}_1$ and $\bar{Y}_2$ the means of the survey variable for the two parts. $\bar{Y} = W_1\bar{Y}_1 + W_2\bar{Y}_2$ is the population mean under the conditions of the survey. Four cases are described below.

(1) The use of $\bar{Y}_1$ to estimate $\bar{Y}$. The bias in this case is $(\bar{Y}_1 - \bar{Y})$ which is equal to $W_2(\bar{Y}_1 - \bar{Y}_2)$ and therefore depends on two factors, W_2, the proportion of nonresponse and $(\bar{Y}_1 - \bar{Y}_2)$, the difference between the means for respondents and nonrespondents. Using $\bar{Y}_1$ as an estimate is equivalent to allocating to the nonrespondents the mean value of the respondents.

(2) The estimation of $Y = N\bar{Y}$, the population total. There are two alternatives. The first is to use $N_1\bar{Y}_1$ which for a sample corresponds to using the simple expansion Fy (Kish, Section 13.4). This has a bias equal to $-N_2\bar{Y}_2$ which depends on the size of the nonresponse group and its mean. It is equivalent to assuming that all the nonrespondents have a zero score on the variable. The alternative is to use $N\bar{Y}_1$, which has a bias of $N_2(\bar{Y}_1 - \bar{Y}_2)$ and which is therefore equivalent to assuming that the means of the respondents and nonrespondents are equal.

(3) The comparison of the means of two subclasses. Frequently, though not always, such differences between means will be relatively free from bias even if the separate estimates are biassed. This will not hold if either the proportion of nonresponse differs between the two classes or if the separate effects of nonresponse are different.

(4) The estimation of more complex statistics. The considerations involved in estimating the effect of nonresponse are based on the assumptions which are implied by each method of estimation. For example, an estimate of a correlation coefficient based on the respondents will be free from bias if the relationship between the two variables is the same for respondents and nonrespondents regardless of differences in their mean values. On the other hand, equality of means will not imply freedom from bias.

The estimation of effects

The results of a survey will invariably include estimates of the effect of nonresponse, whether these are implicit or explicit. The typical situation is one in which the extent of the nonresponse is reported and the results are calculated on the basis of the responses. This approach implicitly assumes that the nonrespondents are sufficiently similar to the respondents to justify ignoring them in the calculations. It is equivalent to treating the achieved respondents as a representative sample from the population. Other explicit procedures are available however and three of them are described below.

(1) Extreme assumptions. This procedure involves making the most extreme possible assumptions about the magnitude of the bias, thus obtaining upper and lower limits for the estimate obtained. Consider the estimation of a proportion, P, based on N units. Observations are obtained for N_1 units of whom P_1 have the characteristic. The maximum value possible for P once we have observed N_1 units will occur if all the nonrespondents possess the characteristic. This is the case where P_2, the proportion for the nonrespondents, is equal to 1. Hence it is safe, although conservative, to take $P_2 = 1$ in calculating the upper limit for P:

$$P_u = W_1 P_1 + W_2 \tag{1}$$

The minimum possible value for P will occur if none of the nonrespondents have the characteristic. In this case $P_2 = 0$. The lower limit is therefore:

$$P_L = W_1 P_1 + 0 \tag{2}$$

Hence to be absolutely safe, the limits of our estimate, taking no account of sampling errors will be W_1P_1, and $W_1P_1 + W_2$. Cochran (1963, p.358) shows that these limits are impracticably large and indeed such conclusions cannot be accommodated in ordinary surveys where nonresponse rates of 10 to 20 per cent are frequent.

(2) Reweighting. A more satisfactory approach is to find out as much information about the nonrespondents as possible and to use this to adjust or reweight the results. Some examples are given in Moser and Kalton (1971). This procedure can be hazardous and we can never be completely confident about the accuracy of the results. It is always desirable to use as much information as can be obtained. Adequate contact sheets on which the interviewer can record demographic and other details are very important. It is also necessary to transfer this information to the data set to be used in the analysis. The methods of data imputation discussed in Chapter 2 are also relevant here.

(3) A successive stages model. This method assumes the existence of a functional relationship between the difficulty of obtaining a response (measured by the call on which the observation is obtained) and the survey variable. Extrapolation beyond the number of calls actually carried out is then possible. An example is given in Hochstim (1967) where the results of successive stages of a survey are reported. The respondents were asked whether (i) they had heard of and (ii) they had had the Papanicolaou cancer detection test. The results are reproduced in Table 1.2.

The figures in Table 1.2 show a clear downward trend in the percentages in both columns at successive stages. An extrapolation of a linear function using the method of least squares to adjust for nonresponses suggests that a better estimate of the proportion of women who had heard of the test would be 77 per cent rather than the 79 per cent finally obtained.

It is worthwhile classifying the results on the survey variable in this way and if a pattern emerges, adjustment may be desirable. Scott (1961) presents results for a number of mail surveys. It is important to realize the assumptions which underly this model. It is implicitly assumed that the nonrespondents are more similar to the

Table 1.2. Results of successive stages of data collection (Reproduced with permission from J. R. Hochstim, *Journal of American Statistical Association*, **62**, 989 (1967))

Stages	No. of women responding	Percentage who had heard of test	Percentage who had had test
One letter	277	88	61
Two letters	381	85	58
Three letters	439	82	55
Telephone	482	82	54
One personal call	501	81	53
Two or more personal calls	524	79	52

late respondents than to the other respondents and this need not necessarily be the case.

In this section nonresponse is discussed from the point of view of its effect on the results once it has occurred. There are many methods for minimizing the level of nonresponse at the stage of the design and execution of the survey. To some extent the problem is insoluble once the fieldwork is over. However the assumptions involved in calculating estimates from an incomplete response should always be made explicit in presenting the results. It is only then that the user of these results can ascertain for himself the likely effect of the assumptions.

1.3.3 Response Errors

It is important to recognize that if a complete enumeration were carried out and even is no nonresponse occurred serious errors might be present in the results due to problems arising from the questionnaire, in the execution of the fieldwork, or from many other causes. The nature and extent of such errors is the concern not only of survey design but also of survey analysis. Response errors will occur in a variety of ways. The respondent may fail to understand a question and thus give an inaccurate answer. Alternatively the respondent may given an answer which he feels will increase his prestige, or may deliberately give an inaccurate response in order 'to please the interviewer'. The errors may arise equally from the interviewer. The manner in which the interviewer asks the questions or the form in which he records the answers may affect the results. A detailed treatment of response errors is given in Chapter 7 of Volume 2. In this section only a brief summary is given.

The expected survey value

Hansen, Hurwitz and Madow (1953) define the *expected survey value* as the result obtained from a complete enumeration of a population based upon any specified set of procedures for obtaining the information. The expected survey value will vary therefore with the particular approach used to obtain the information. The results will be determined by the actual procedures adopted – the training and supervision of the interviewers, the wording of the questions, the circumstances under which the information is collected, and many other factors. The difference between the expected survey value and the 'true' value which is being estimated is called the *response bias*. The second component of response errors can be described as *response variance*. This is a measure of the effect of those variable errors whose absolute effects have an expected value of zero.

Measurement of response errors

In order to obtain a measure of response bias it is necessary to have information from outside the survey itself. The estimate requires the existence of data which measures or estimates the true value more accurately than does the survey. Such information may frequently not be available or may be difficult and expensive to obtain. It is considerably easier to estimate response variance by suitable design of the study. The effect of different wordings of questions or of different questionnaires may be estimated by allocating them to random subsamples of the population. If the variation between different interviews is to be measured this may be accomplished by the allocation of a random sample of the interviews to each interviewer. In the analysis the variation between the results obtained by the interviewers can be compared with the variation to be expected if the interviewers all did uniformly good (or bad) work. This measures variation about the expected value and takes no account of response bias.

It is important to realize that response errors cannot be reduced or eliminated simply by increasing the number of observations. They apply equally to complete enumerations and sample surveys. Many of the problems can be tackled by appropriate survey design. The reader is referred to Hausen, Hurwitz and Madow (1953), Kish (1965), Moser and Kalton (1971) and Chapter 7 of Volume 2. In the case of interview surveys, response errors can be intrinsically interesting however and may be seen as the outcome of a controlled interaction between the interviewer and the respondent. Thus information about the survey variable may be extracted from the pattern of response errors. This and other analytical aspects of response errors are described in Chapter 7 of Volume 2.

1.4 THE SAMPLE DESIGN

Sample design has two aspects, the selection process which is the procedure by which elements from the population are selected for inclusion in the sample, and the estimation process which is the method or operations used to compute the sample statistics with which we estimate the population parameters. This is the area of survey analysis most adequately covered in the literature. Beginning with Yates (1949), numerous textbooks have discussed in detail the problems both of selection and estimation. Cochran (1963) gives a technical presentation of the principal results and derivations. The most useful book for the survey practitioner is Kish (1965) which not only presents the principal results and their derivations but also raises and gives solutions for the practical problems which arise in sample design and analysis. This section draws heavily on this book and on the paper by Kish and Frankel (1974).

The sample survey is now commonly accepted as a respectable method of social inquiry. A brief comparison of sample surveys and complete censuses is appropriate here. The purpose of *survey sampling* is to select and observe some part of the population and on the basis of this selected sample to make inferences about the population from which it is drawn. A complete census may have some advantages over a sample survey. First, it is possible to obtain and analyse data for small segments of the population : towns, villages and districts, for example. Second, the results will probably be more acceptable to the user because they are based on complete coverage. Third, the bias due to noncoverage may be easier to check and reduce. On the other hand sample surveys also have some advantages. First, they are more economical. Second, they can be completed more quickly and consequently are more suitable if results are required urgently. Third, they may be more accurate than complete coverage, since the number of skilled personnel necessary for data collection for a complete enumeration may not exist. It is possible to view a complete census as being equivalent to a sample. To the extent that response errors occur in the investigation, the expected survey value obtained from a complete coverage is only one of many possible values which could have arisen from essentially the same operation. Also, in examining social processes the complete enumeration of a population at a point in time may be less efficient than the selection and observation of smaller samples taken at different times. The universe to which social research results are generalized is wider, usually in extent and certainly in time, than the population being studied.

1.4.1. Selection procedures

Model Sampling

The term *model sampling* is used by Kish to describe the class of selection procedures which are based on broad assumptions (usually not explicit) about the population

The first category is *haphazard sampling* which is any procedure which selects into the sample any population elements which happen to be available. This includes samples of volunteer subjects so widely used in psychology. This method, in common with other model sampling methods, has the disadvantage that the population to which inferences can be made is not clearly defined. The basic assumption is that the elements or individuals included are in some way representative of the population; but the justification of this assumption is impossible by the nature of the selection procedure.

A second procedure can be described as *purposive sampling*. This involves the use of personal judgement of the investigator in selecting 'representative' elements. Again rigorous inferences cannot be made to the population from which the sample is selected but only to some hypothetical population of which the sample may be representative.

Quota sampling is a generic term which describes loosely a set of sampling methods used widely by market research organizations in interview surveys. The distinguishing feature of a quota sampling scheme is that each interviewer's workload is based on a fixed quota of individuals who must be representative of the population in respect of some characteristics. The characteristics usually employed are age, sex and socioeconomic status. For example, if an interviewer has 100 interviews to complete, the proportion of males in the sample must be equal to the proportion of males in the population; the proportion of respondents over 65 must be the same as the proportion of the population over 65, etc. Given these restrictions, which make up the *quota controls*, the interviewers then do their best to find people who satisfy the requirements.

Quota sampling has the characteristic, common to all non-probability sampling, that the sampling procedure is ill-defined. There is no way of knowing what chance of selection any individual in the population has had. Within the restrictions imposed by the quotas, the choice of respondent is left entirely to the discretion of the interviewer. This dependence of the achieved sample on the (non-measurable) whim of the interviewer is a source of bias in the selection procedure and thus can give rise to a bias in the estimate derived from the survey.

Quota sampling is practised mainly because its cost per element is lower than for probability sampling. However in terms of the precision of the estimate achieved this reduction in cost may be spurious. The less restrictive the quota controls, and therefore the more scope for selection bias on the part of the interviewer, the cheaper quota sampling becomes. A reduction in monetary cost at the price of increased bias is not necessarily a good bargain!

The most important criticism of all forms of model sampling, however, is paradoxical. There is no such thing as a 'representative', 'fair', or 'unbiased' sample. These terms apply only to the procedure by which a sample is drawn or to the expected value of the estimators over repeated trials and not to the sample itself.

Only by a knowledge of how the estimate derived from the sample can vary from one possible outcome of the sampling procedure to another can the procedure be judged. In the case of quota sampling, since the method of selecting the sample is not clearly defined, there is no way in which the likely accuracy of the estimate can be discovered. Due to the selection bias, some individuals may have no chance of selection; but there is no way of knowing which individuals these are. Therefore not only is there a non-measurable bias but also the possible fluctuations from one sample to another remain unknown.

Despite their faults, model sampling procedures do sometimes produce good results. The fundamental problem however remains that due to the nature of the selection procedures involved, it is not in general possible to define the population to which inferences can properly be made unless the untestable implicit assumptions can be justified. Thus the whole body of statistical methodology which is based on the ideal of probability sampling of elements is not appropriate to such samples.

Probability sampling

A probability sample is selected by the use of some probability mechanism (for example, tables of random numbers) and the *probability of selection into the sample of each element in the population must be known and non-zero*. The term *random sample* is sometimes used but can be confused in non-technical terminology with haphazard. Most probability samples are designed to be *measurable*, which implies that the design is such that the precision of the estimator may also be obtained from the sample observations. Kish (1965) presents a taxonomy of probability selection procedures which is reproduced in Table 1.3. The term *epsem* is derived from the words 'equal probability of selection methods' and includes all samples in which the overall probability of selection for each element in the population is the same. Both epsem samples and unequal probability samples can arise through any of the other procedures given in the table. The great advantage of epsem samples is that the observations need not be weighted in the analysis,

Table 1.3. A taxonomy of probability selection methods. Five alternatives that may be combined (Reproduced from L. Kish, *Survey Sampling*, John Wiley and Sons Inc., New York, p. 20)

I *Epsem*: equal probability for all elements (a) Equal probabilities at all stages (b) Equal overall probabilities for all elements obtained through compensating unequal probabilities at several stages	*Unequal probabilities* for different elements; ordinarily compensated with inverse weights (a) Caused by irregularities in selection frames and procedures. (b) Disproportionate allocation designed for optimum allocation
II *Element Sampling*: single stage, sampling unit contains only one element	*Cluster Sampling*: sampling units are clusters of elements (a) One-stage cluster sampling (b) Subsampling or multistage sampling (c) Equal clusters (d) Unequal clusters
III *Unstratified Selection*: sampling units selected from entire population	*Stratified Sampling*: separated selections from partitions, or strata, of population
IV *Random Selection* of individual sampling units from entire stratum or population	*Systematic Selection* of sampling units with selection interval applied to list
V *One-Phase Sampling*: final sample selected directly from entire population	*Two-Phase (or Double) Sampling*: final sample selected from first-phase sample, which obtains information for stratification or estimation

whereas with unequal probability samples, weights inversely proportional to the probability of selection must normally be used.

Element sampling covers all the situations in which the population elements are also the sampling units. The basic element sampling design is *simple random sampling*. A simple random sample is a sampling scheme in which not only does every element in the population have an equal probability of selection, but also each combination of n elements has the same probability of selection. This is the fundamental sampling scheme on which the results of classical statistical inference are based. However in practice departures from random sampling are either desirable or necessary. *Cluster sampling* involves the selection of groups of elements together. The conditions which favour cluster sampling are: (i) that the clusters should be easily and uniquely defined and identified; (ii) that once a sample element has been identified, collection costs are relatively low for additional neighbouring elements; (iii) that the elements in a cluster are sufficiently dissimilar to justify the inclusion of additional elements beyond the first. Frequently, cluster sampling requires a larger sample size than simple random sampling to provide estimates with a specified level of precision. Therefore, if the decision to select a cluster sample is to be justified, the reduction in the cost of sampling and data collection must be large enough to compensate for the additional costs of processing and analysing a more complex sample. There are situations where cluster sampling may be possible although element sampling may be ruled out because of practical difficulties. For example, in sampling workers from manufacturing industry, it is possible to obtain a list of establishments but a list of all workers may be out of the question. Once the sample of eatablishments has been selected lists may be obtained within each from which a sample of workers can be drawn. In such a case a two-stage cluster sample may be the only possibility. In fact most surveys of large populations are multi-stage samples, the first-stage units or clusters typically being districts, constituencies or counties, Only in the case of compact populations is element sampling employed.

Stratification implies the division of the population into a number of subgroups or strata. An attempt is normally made to ensure that each subgroup or stratum in the population is constructed in such a way that the individuals or elements within a stratum resemble one another. In stratified random sampling an independent random sample is selected from within each stratum.

For example, if we wish to estimate the attitudes of adults to the Women's Liberation movement, we might *stratify by sex*, i.e. categorize the population by males and females and select a sample from each group, rather than a single sample from the whole population. Since we might expect that men and women would have different attitudes to Women's Lib, we ensure by stratifying that both men and women are adequately represented in the sample, and therefore we improve the precision of our estimator.

A second example may help to reinforce this point. Before an election the opinion pollsters wish to predict shares of the vote. To do this they select a small number of constituencies in which to carry out interviews. It is essential if the prediction is to be accurate that certain sections of the community should be represented in the sample, e.g. (i) rural areas and urban areas; (ii) Labour strongholds and Conservative strongholds; (iii) marginal constituencies. Therefore before selecting the consitiuencies in which the interviews are to be carried out the full set of constituencies is divided into these categories and a separate sample is selected from each category, ensuring that each is adequately represented. This procedure should increase the precision of the estimator.

The examples given above illustrate a second advantage of stratification: it increases the acceptability of the results of a sample survey since it can be shown that certain key divisions of the population are represented in the sample.

The third advantage of stratification is that for large scale surveys it can be administratively convenient. For example in a national survey it is essential that arrangements for distribution of fieldwork materials, training and briefing of interviewers, etc. should be made well in advance. *Stratification by region* will ensure that this is possible since it can be predicted accurately how many interviews will be required in, for instance, the North-West region.

The fourth advantage of stratification arises from the fact that the sample is selected independently in each stratum; thus the method by which the sample is selected can be varied from one stratum to another, as appropriate. It is useful to distinguish between two purposes for stratification: i.e. stratification for the comparison of *domains of study* and stratification to increase the precision of the overall estimator. It is necessary to have prior knowledge of a relationship between the survey variable and those available as stratifiers. Each element must be assignable to a particular stratum before sampling. A problem arises if some of the elements are misclassified. If the purpose of the stratification is to increase overall precision the only result is that the gain in precision is reduced. However, if the purpose is to compare the strata, then adjustments must be made to take into account the fact that the strata no longer represent exactly the domains of study (categories of the population) we wish to compare.

Proportionate stratified sampling in which equal probabilities of selection are applied in each stratum will be more precise than simple random sampling if the mean of the survey variable differs from one stratum to another. *Disproportionate stratified sampling* will be still more efficient if a larger sampling fraction is applied to the more variable strata. However the danger of disproportionate allocation is that if the estimates of variability on which the allocation is based are wrong the loss of precision may be severe.

Systematic selection is a convenient and frequently used practical procedure. It avoids the necessity of drawing many random numbers and can be used to provide further stratification. For a population with a linear trend in the listing the precision of a systematic sample will be higher than that of a simple random sample but lower than that of a stratified sample (Cochran, 1963). Systematic sampling can suffer from periodic fluctuations in the population list but this is not a common problem. If the periodic structure of the population is known, a systematic sample can usually be designed to capitalize on it. Otherwise, if a periodic effect is suspected but not known, a simple random or stratified random sample is preferable.

Multiphase sampling is particularly powerful in some circumstances:

> 'It is sometimes convenient and economical to collect certain items of information on the whole of the units of a sample, and other items of information on only some of these units, these latter units being so chosen as to constitute a subsample of the units of the original sample. This may be termed two-phase sampling. Information collected at the second or subsampling phase may be collected at a later time, and in this event, information obtained on all the units of the first-phase sample may be utilized, if this appears advantageous, in the selection of the second-phase sample. . . . .
>
> 'An important application of multi-phase sampling is the use of information obtained at the first phase as supplementary information to provide more accurate estimates (by the method of regression or ratios) of the means, totals, etc., of the variates obtained only in the second phase.' (United Nations, 1950).

Two-phase sampling is sometimes called double sampling. The basic idea involves first selecting a large sample of size n_L from the population and selecting the final sample of size n from this larger sample using information collected from the preliminary sample. It is in essence a compromise between the detailed observation of a small sample and the superficial examination of a large sample. The first phase may also be used as a screening stage at which eligible members of the population are identified.

The five component parts of Table 3 may be combined one with another and many sample designs will incorporate more than one of the components. The detailed treatment of estimation from complex sample designs is not possible here and is dealt with for example in Kish (1965). In Section 4.2 the principal features and problems of estimation for some common sample designs are discussed.

1.4.2 Estimation

The assumption underlying standard statistical inference is that of simple random sampling. However in survey sampling two special features are present. First, the population from which the sample is drawn is finite. Second, the sample designs in common use are considerably more complex than simple random sampling. The principal departures from simple random sampling are due to stratification and clustering. The principles of inference are discussed in Chapter 1 of Volume 2 for the case of simple random sampling, assuming independence between the observations. These cover the estimation of complex analytical statistics as well as the simpler case of the estimation of means and totals. Survey sampling theory has been developed to deal with the problems of estimation for the complex sample designs used in social surveys. The scope of the theory is indicated in five major books published some twenty years ago: Yates (1949), Deming (1950), Cochran (1953), Hansen, Hurwitz and Madow (1953) and Sukhatme (1954).

The treatment in those books concentrates on providing an estimator of the population mean $\bar{Y}$ and an estimate of the precision of this estimator. The precision is presented in the form of a standard error and is used to provide interval estimators $\bar{y} \pm t_p$ se $(\bar{y})$ for the population parameter $\bar{Y}$. Such estimators are based on the sample design and take into account its complexities. The approach however

Table 1.4. The present status of sampling errors. Row A is the domain of standard statistical theory, and column 1 of survey sampling. (Reproduced with permission from L. Kish and M. R. Frankel, *Journal of the Royal Statistical Society, Series B,* **36**, Part 1, Figure 1 (1974))

Selection methods	1 Means and totals	2 Subclass means and differences	3 Complex analytical statistics, e.g. coefficients in regression
A. Simple random selection of elements			
B. Stratified selection of elements		Available	Conjectured
C. Complex cluster sampling		Available	Difficult: BRR, JRR, TAYLOR

is not convenient for the computation of large numbers of standard errors required in a social survey nor does it encompass the increasingly common situation where complex statistics such as regression and correlation coefficients are calculated from the sample observations.

Table 1.4 reproduces a table from Kish and Frankel (1974) which divides into three levels of complexity both the selection methods and the statistical estimators. The division, although arbitrary, is meaningful and useful. It covers the range of estimation problems common to most social surveys. The shaded portion indicates the areas where good information and techniques are available and documented. The remaining cells still present some problems.

The principles of estimation for row A of the table are dealt with in Chapter 1 of Volume 2. Column 1 is the area which is covered in most textbooks on survey sampling. The most important categories for the survey analyst are those in rows 2 and 3.

Simple random sampling

There are three basic results in sampling without replacement from a finite population. For simplicity, the results are confined to the estimation of a population mean $\bar{Y}$.

Let N be the population size; n the sample size

$\bar{Y}$ be the population mean; $\bar{y}$ the sample mean

$S^2 = \Sigma(Y_i - \bar{Y})^2/(N-1)$; $s^2 = \Sigma(y_i - \bar{y})^2/(n-1)$

Then

$$E(\bar{y}) = \bar{Y} \tag{1}$$

i.e. the expected value of the sample mean over all possible samples of size n is equal to the population mean; the sample mean is an *unbiased estimator* of the population mean.

$$\text{Var}(\bar{y}) = \left(1 - \frac{n}{N}\right)\frac{S^2}{n} \tag{2}$$

where $\text{Var}(\bar{y})$ denotes the sampling variance of $\bar{y}$.

$$E(s^2) = S^2 \tag{3}$$

If

$$\hat{\text{V}}\text{ar}(\bar{y}) = \left(1 - \frac{n}{N}\right)\frac{s^2}{n} \quad \text{then,} \quad E(\hat{\text{V}}\text{ar}(\bar{y})) = \text{Var}(\bar{y}). \tag{4}$$

Consider equation (2): the variance of the sample mean depends on three factors. First, the sampling fraction, $f = n/N$. The effect of the sampling fraction appears in the term $(1 - f)$ which is called the *finite population correction*. In most sample surveys where the population is very large the value of f will be small and the effect of the finite population correction will be slight. Second, the variability S^2, in the parent population. The more variable or dispersed the parent population, the larger the value of $V(\bar{y})$, *ceteris paribus*. Third, the sample size n. This is the crucial factor. It is the absolute size of the sample and not, in general, the sampling fraction which is most important in the determination of the variability of the sample estimators. In practice the value of S^2 is not known. However an unbiased estimator of S^2 can be obtained from the sample (3). Consequently the variance of the sample can also be estimated directly from the sample (4). If n is large ($n \geqslant 50$, say) and $(N - n)$ is also large, we can appeal to the finite population theory for simple random sampling (Erdös and Renyi, 1959 and Hájek, 1960) which states that under certain conditions, usually obtained in practice, that

$$\frac{\sqrt{n}(\bar{y} - \bar{Y})}{\sqrt{(1 - f)} \, . \, s}$$

tends to the normal distribution with mean zero and variance one as both n and $N - n$ tend to infinity. Thus we can set approximate confidence limits for reasonably large samples, e.g.

$$\bar{y} \pm 1.96 \frac{\sqrt{(1 - f)} \, . \, s}{\sqrt{n}}$$

is an approximate 95 per cent confidence interval for $\bar{Y}$.

In general,

$$\bar{y} - t(\alpha, \infty) \sqrt{\left\{\frac{(1 - f)}{n}\right\}} \, . \, S \leqslant \bar{Y} \leqslant \bar{y} + t(\alpha, \infty) \sqrt{\left\{\frac{(1 - f)}{n}\right\}} \, . \, S \qquad (5)$$

holds on an average with a probability of $(1 - \alpha)$, where $t(\alpha,\infty)$ is the value of the normal variate corresponding to the value of $(1 - \alpha/2)$ of the normal probability integral. This provides the basis for all inference from survey samples and it is an extremely powerful result. It states that not only can an unbiased estimator of the population mean be obtained from the sample but the sample itself also provides us with sufficient information to obtain an unbiased estimator of the variance (or precision) of this estimator.

The results above are so simple and easy to apply that there would seem to be no excuse for failure to present estimates of sampling variance (*sampling errors*) together with the estimators derived from a sample survey. The large numbers of estimates produced from a survey however, together with the complexities both of design and estimation indicated in Table 1.4 above present real difficulties for the survey practitioner.

The Design Effect

A single key concept permits the discussion of sampling errors in general in terms of the results for simple random sampling given above. The design effect (*deff*) is 'the ratio of the actual variance of a sample to the variance of a simple random sample of the same size' (Kish, 1965, Section 8.2). Thus,

$$\text{deff} = \frac{\text{Var}(\bar{y})}{(1-f)\dfrac{S^2}{n}} \tag{6}$$

This measure makes it possible to compare the effects of different departures from simple random sampling on the precision of the estimators derived from a survey. The discussion below gives a brief summary of the principal conclusions. For specific formulae and discussion the reader is referred to Kish (1965) and Kish and Frankel (1974).

Stratification [Row B, Table 1.4]

The effect of stratification on the precision of estimators depends on the efficiency of the stratification, that is, the extent to which the elements differ from one stratum to another. Proportionate allocation of elements (or equal sampling fractions within strata) is the most common allocation used in practice. Usually only small gains in precision will result from the proportionate stratification (i.e. gains of the order of 5 to 10 per cent) because the variables which are commonly available for stratification (such as region and sex) are not closely related to the survey variable and do not divide the population into homogeneous groups. However the use of stratification is justified by the gain in precision together with the fact that it is an epsem procedure which results in using unweighted means in the analysis. In terms of the design effect, the position can be stated as follows: *for proportionate stratified element sampling the design effect is normally less than 1.*

In most surveys, analysis is also carried out for subclasses of the sample, e.g. for all men and women separately, for different age groups, etc. The term *crossclasses* was coined by Kish to describe subclasses which cut across the strata used in the selection of the sample. An important result applies here. For the means of crossclasses the gains of proportionate stratification decrease as the proportion of the crossclass decreases. For small crossclasses the gains tend to vanish and the variance approaches that of simple random sampling. In other words *the design effect tends to 1 from below for crossclasses as these become small* (Kish and Frankel, 1974).

Another important case is the comparison of crossclasses. Here again the result is encouragingly simple. For differences between the means of crossclasses the gains of proportionate stratification tend to vanish. *The design effect for the comparison of crossclasses is close to 1 in this case.*

Using these results as a basis, we can conjecture that for more complex statistics computed for stratified element samples a similar situation will hold. This conjecture is supported by the data of Nathan (1972) on chi-squared tests. In general the design effect for analytical statistics may be expected to be close to 1.

The position for proportionate stratification can easily be summarized. The design effect will be close to, but generally less than, 1 for all sample statistics based on stratified element sampling.

A special case occurs when the values of the stratification variable are not known for each element in the population in advance but when accurate information is available on the number of elements which belong to each category of the stratification variable. Examples will occur in terms of the age and sex distrubution of the population in national surveys for instance where up-to-date national figures may be available, whereas information on sex and age will not appear on the sampling frame for the elements. In such cases the technique of stratification after selection or *post-stratification* may be used in the analysis. Suppose a simple random sample is drawn and the units are classified into strata after looking at the observations. Let $n_1, n_2, \ldots, n_L$ be the numbers of observations in the respective strata. We can form the estimator

$$\bar{y}_{st} = \sum_{h=1}^{L} W_h \bar{y}_h, \tag{7}$$

where there are L strata, W_h is the population proportion which falls in the hth stratum, and $\bar{y}_h$ is the mean of the sample observations in the hth stratum. The variance of this estimator is calculated conditional on the achieved sample numbers in the strata. To assess the precision of this method we want the average value of the variance over repetitions of the whole procedure.

This can cause difficulties since in extreme cases some values of n_h could be zero (no sample elements appearing from stratum h) but the probability is small when the sample size n is large.

It can be shown, that when n is large, the precision of post-stratification is almost as high as that of proportionate stratification. *The design effect will still be less than 1 for post-stratification*. This is a useful analytic technique which can be applied to survey results even when stratification is not included in the sample design. It can also be applied to subclasses.

Disproportionate stratification or different sampling fractions in different strata is less commonly used. It can be extremely powerful in some circumstances but the information necessary to apply it successfully is not generally available. Besides, errors in the allocation can lead to a serious loss in precision. In most multipurpose surveys it is not appropriate (Kish, 1961).

When the survey variable is in the form of a proportion the gains of stratification are even more modest, and disproportionate allocation is seldom worth the resulting complexity of dealing with weighted estimators.

Cluster Sampling [Row C, Table 1.4]

The effect of cluster sampling on the precision of the estimators can be described most easily in the case of simple random selection of complete equal-sized clusters.

Let the population of N elements consist of A clusters with B elements in each. Consider a simple random sample of a complete clusters. Thus $n = aB$ and $a/A = f$.

Let

$$\bar{Y}_\alpha = \frac{1}{B} \sum_{\beta=1}^{B} Y_{\alpha\beta}$$

be the mean of cluster α and $\bar{Y}$ be the population mean. Then, if

$$\bar{y} = 1/a \sum_{\alpha=1}^{a} \bar{Y}_\alpha,$$

$$E(\bar{y}) = \bar{Y} \tag{8}$$

and the variance of this estimator can be written as

$$\text{Var}(\bar{y}) = (1 - a/A)\frac{S_a^2}{a} \tag{9}$$

where S_a^2 is the variance between the cluster means in the population. However to help appreciate the meaning of (9) the variance can be written as

$$\text{Var}(\bar{y}) \doteq (1 - f)\frac{S^2}{n}\,[1 + \rho(B - 1)] \tag{10}$$

where S^2 is the element variance in the population, and ρ is the *intra-cluster correlation coefficient*. This coefficient is a measure of the internal homogeneity of the clusters. The intra-class correlation coefficient gives an indication of the dissimilarity between elements within the cluster compared to the dissimilarity between elements within the population. The more similar the elements within a cluster are to one another, the larger the value of ρ will be. Utilizing the definition of the design effect (6) it follows directly from (10) that the design effect for a sample of complete clusters is

$$\text{deff} \doteq 1 + \rho(B - 1) \tag{11}$$

Thus the precision of a cluster sample will depend on two factors: the intra-cluster correlation coefficient ρ and the cluster size B. For two-stage sampling where b elements are selected from each cluster, $n = ab$, a similar result holds:

$$\text{deff} = 1 + \rho(b - 1) \tag{12}$$

These results suggest that a coefficient of homogeneity could be defined for the more complex and common case of unequal size clusters or unequal subsample sizes within clusters. Kish (1965) and Kish, Groves and Krotki (1975) suggest the use of a *synthetic* intra-cluster correlation coefficient, roh, which is computed as

$$\text{roh} = \frac{(\text{deff} - 1)}{\bar{b} - 1} \tag{13}$$

where $\bar{b}$ is the average subsample size within cluster. These values of roh measure the degree to which the values for variables are homogeneous within sample clusters.

The intra-cluster correlation coefficient is usually positive and in these cases the design effect is greater than 1. The highest possible value of roh is +1 and this corresponds to complete uniformity within clusters. In this case the design effect is equal to b. The minimum possible value of roh is $-1/(b-1)$ which corresponds to the case where the cluster means are all equal. It is unusual in practice to find cases where roh is negative. Consequently, *in general for cluster samples, the design effect is greater than 1.* In the case of complex statistics including the comparison of means of crossclasses, design effects will also tend to be greater than 1. Standard errors computed on the assumption of simple random sampling of elements will be underestimates of the true standard errors. The degree of underestimation may be severe, since design effects as high as 10 have been reported. The *design effect for subclass means tends linearly to 1 as the size of the subclass decreases* if ρ is constant over different sizes of subclasses (Kalton and Blunden, 1973). The design effects for complex statistics are generally lower than those for the means of the same variables. This is important since the design effects for means are easier to compute and can therefore be used as conservative estimates for the design effects of the complex statistics. Kish and Frankel (1974) present results which show that if the design effects for means are high then the design effect for complex statistics based on these variables will also be high.

Computation of Sampling Errors

The results above provide a set of general guidelines to the effect of the sample design of the estimators derived from a survey. However no magic numbers are available which remove the necessity of calculating the sampling errors in each particular case.

(i) Replicated Sampling. The idea of replicated sampling is that the full sample is divided into c parts in each of which the same sample design is used. The usefulness of the procedure therefore depends on the appropriateness of the sample with which the analyst is working.

Let x_γ be the sample statistic based on replicate γ, $\gamma = 1, \ldots, c$. The sample statistic for the full sample can be computed as $\bar{x} = \Sigma x_\gamma / c$. If the replicates are independent then the variance of $\bar{x}$ can be estimated by

$$\text{Var}(\bar{x}) = (1-f) \sum_{\gamma=1}^{c} \frac{(x_\gamma - \bar{x})^2}{c(c-1)} \tag{14}$$

In the case of the estimation of the sample mean the x_γ will be the means of the sample replicates and $\bar{x}$ will be equal to the overall sample mean. In the case of statistics such as the regression coefficient $\bar{x}$ will not be equal to the value of the statistic computed for the sample as a whole. In this situation we would use as our estimator the full sample value but estimate its variance using (14).

This is a supremely flexible approach to variance estimation. It uses the simple idea of estimating the precision of the sample statistic in terms of the variability between subsamples within the sample. It is not necessary to take account explicitly of the complexity of the sample design since the formulation includes it implicitly. There is one problem however. Many samples are not designed as a set of independent replicates and the method therefore is not always appropriate.

(ii) Design-specific computation. In a survey where the number of estimators being produced is small and the sample design is not too complex it is possible to calculate the sampling errors directly using the formulae presented in the survey sampling textbooks.

(iii) Computer programs. The need for efficient computer programs has been expressed succinctly by Kish (1971):

> 'The theory of survey sampling deals largely and adequately with the estimation of variances for sample means and aggregates based on designed samples. However that theory does not satisfy the needs of survey practitioners. First, the aims and results of most surveys cannot be expressed in one or two statistics, but in scores, hundreds or thousands – too many for the standard methods of computing and presenting sampling errors. Second, the needed statistics are usually not ordinary sample means, but the means and differences of subclasses which cut across the sample designs. Third, survey research results are increasingly often expressed in multiple regressions and other analytical statistics.
>
> This divorce of theory and practice has illegitimate results. Most frequently the sampling errors of survey statistics are not computed, or they are presented with simple random sampling assumptions; this is sometimes improved with a "borrowed" factor for the design effect.'

Three programs have been developed by the Institute for Social Research at the University of Michigan and make up the *Sampling Error Program Package* [SEPP, Kish, Frankel and Van Eck, 1972].

The first of the programs is ABSERD (A Basic Program for Sampling Errors of Ratios and Differences) and is useful if only a few results are needed. The second, PSALMS (Paired Selection Algorithms for Multiple Subclasses) is far more powerful and can deal rapidly and economically with many statistics in a single run. A large computer is however needed. The third, BRRP (Balanced Repeated Replication Program) is designed mainly for the computation of the sampling errors of multiple regression coefficients and other analytical statistics. It is a more complex program that either of the others and although it can also deal with similar problems it is more expensive to use. It can however cope with problems not dealt with by ABSERD and PSALMS.

The three programs together can handle almost all the computations of sampling errors necessary in survey analysis.

1.4.3. Summary

The sample design plays a key role in the analysis of survey data. Both the selection process (Section 4.1) and the estimation process (Section 4.2) are important. The survey analyst is frequently faced with the selection process as a *fait accompli* but an understanding of the principles of selection is important in determining the likely effect of the selection process on the analysis. Section 4.2

summarizes the effects of departures from simple random sampling in terms of the design effect. These results provide some general indications of the kind of results which can be expected. They are however no substitute for the calculation on sampling errors for a particular survey. The importance of the calculation and presentation of sampling errors in the analysis of survey data cannot be overemphasized. The basic theory underlying the estimates of sampling errors is well established and computer programs are available for their computation. The use of point estimates of parameters or the use of the simple random sampling variance cannot be justified for surveys with complex sample designs.

1.5. Methods of Analysis

A consideration of the appropriateness of tests of significance in social research is a useful starting point for a discussion of methods of analysis. Three papers in the late 1950's present an interesting evaluation of the problem. Selvin (1957) criticizes tests of significance in a paper which goes beyond considering the tests themselves and questions the very nature of social research. His argument centres on the need to (i) control extraneous variables in an experimental way and (ii) randomize the uncontrollable. This paper forms a useful basis for discussion but offers little advice on how to proceed. McGinnis (1958) takes the view that Selvin's paper has nothing to do with the admissibility of tests of significance in social research and he introduces the idea of considering classes of hypotheses. Kish (1959) puts the debate in a practical perspective and considers the use and misuse of tests in a search for causation.

Selvin claims that the conditions under which tests of significance may validly be used are almost impossible to achieve in survey research and that even when those conditions are met the nature of the research situation is such that correct inferences are difficult to make. His concern is not with statistical procedures but with the problems of design for testing hypotheses and the interpretation of such tests. In survey research it is seldom possible to randomize and in most situations a division into experimental and control groups is not applicable since no stimulus or set of treatments is administered to one group and withheld from another. As an example he considers a case where the relationship between political interest and place of residence (urban *vs.* rural) is being examined and asks under which conditions, if any, can the differences be subjected to tests of significance. As there is no random assignment to the two groups, he suggests that it seems beside the point to test.

In an attempt to consider the problem of using tests in a search for causation it is necessary to distinguish between correlation and causation. To clarify the position Kish defines four classes of variables:

I. *Explanatory*. These are the objects of research. They are designated on the basis of substantive scientific theory and are the variables among which the investigator wishes to find and measure relationships. Some of these may be *dependent* variables

and others may be *independent* or *predictor* variables whose effect on the dependent variables we wish to examine.
II. *Extraneous controlled.* The control may be exercised in the design (i.e. in the selection process) or in the analysis (i.e. in the estimation process).
III. *Extraneous uncontrolled.* These may be confounded with some of the variables in class I.
IV. *Extraneous uncontrolled (assumed) randomized.* These are assumed to be random errors. In perfect experiments they are actually randomized whereas in social surveys this randomization is merely assumed.

In the ideal experiment (as described by Selvin) no variables appear in class III. The testing procedure then consists of testing the effects in class I against the randomized errors in class IV. In survey analysis the main problem is the separation of the effects of the variables in classes I and III. The advantages of experiments are that confounding variables are eliminated and that there is control over the introduction and variation of the predictor variables. In addition there is more flexibility for statistical manipulation. It should be recognized however that it is never possible to control all potentially confounding variables, even in an experimental situation. In surveys it is impossible to solve simultaneously all the problems of measurement, representation and control. Probability samples permit inference statements to be made in the 'natural setting' of the population. The principal weakness is the lack of control. But the judicious collection and use in the analysis of ancillary information can provide a considerable degree of control of the major confounding variables in class III.

The purpose of statistical tests of significance is not to point to causation or explanation. They ask basically whether there is anything which requires explanation. The tests tell us nothing about the magnitude and importance of the relationships observed. The function of the tests should be to tell us whether the variation observed is large enough to have some confidence in the result and whether the result is consistent with some assumed relationship of substantive importance.

The level of knowledge in many areas of social research is such that meaningful hypotheses expressible in terms of the magnitude and strength of relationships between variables are difficult or impossible to find. In many instances the research is carried out as an exploration of the field of study. Null hypotheses of no difference are frequently known to be false before the data are collected and the rejection of such hypotheses simply reflects the size of the sample and the power of the tests (see Chapter 1 of Volume 2). Thus what is required is probably description, although the description may be in terms of complex relationships, and some indication of the precision of this description. In some cases there is sufficient prior knowledge to construct a formal model which can be tested. But above all the results of the analysis should comprise a coherent assessment of the magnitudes of the interrelationships found in the data.

1.5.1 Description

The discussion here is restricted to quantitive analysis. The simplest type of analysis consists of the estimation from the sample of means, totals, simple correlation coefficients and frequency distributions. This basic description of the data should be part of the inference from the sample to the population and the measures should be presented together with their sampling errors. This aspect of analysis is not dealt with in detail in this book. The survey sampling textbooks provide more than adequate coverage of this area for the estimation of means and totals. The considerations which apply there also apply no less strongly to the estimation of covariance and correlation. In Chapter 6 of Volume 2 an empirical investigation of the effects of a complex sample design on estimates of correlation is presented. The design chosen is that of a stratified cluster sample and different relationships between the stratification variable and the variable whose correlations are being estimated are considered.

1.5.2 Searching for structure

In most analyses of survey data however the main interest is in considerably more sophisticated analysis than the reporting of descriptive statistics. In many cases the orientation is not towards the testing of specific hypotheses but towards discovering regularities or irregularities in the data once they have been collected. The prior contribution of the substantive theory is in determining which variables are to be observed but not in specifying the precise functional form of the relationships between them. The overall situation with which the survey analyst is faced is a set of observed variables

$$\{y_1, y_2, \ldots, y_m, x_1, x_2, \ldots, x_p\} \tag{15}$$

where the y's are the dependent variables and the x's are a mixture of independent variables. The relative status of the variables is difficult to define. There may not always be any variables which have the special status of *dependent* variables. In some cases the interest lies simply in the interrelationship between a set of variables, where all have equal status. The search for relationships can be classified according to the purpose of the analysis.

The first and simplest purpose of the multivariate analysis may be *data description or data reduction.* The aim in this case is to reduce the volume of data by transforming the full data set into a more compact form which preserves its essential characteristics and which provides an accurate summary. Such a reduced data set may be an end in itself or it may be used as input for further analysis.

One of the principal concerns of many investigators is to discover whether their data reflects some basic *typology* in their field of study i.e. whether well-separated clusters of individuals exist in the population where the clusters can be interpreted meaningfully in the context of the substantive theory. The techniques are usually used in situations where there is little or no prior information about the structure or pattern which may exist amongst the individuals or variables on whom the data are collected. The identification of groups or clusters of individuals or variables which are in some sense internally homogeneous and externally heterogeneous is the principal objective of such analyses. Most of these techniques are essentially non-inferential in the sense that they do not in general have any associated distribution theory or tests of significance.

In many situations the survey variables are merely proxies for the constructs in which the investigator is interested. The variables which are observed may reflect an attempt to span or measure a few underlying dimensions. The *identification of such latent dimensions* from the manifest or observed variables is a further purpose of the analysis of survey data. An example is the case of *index construction* in attitude measurement which represents a particular use of multivariate techniques in survey analysis.

When one or more of the survey variables can be classified as dependent, predictive *model-building* frequently becomes the aim of the analysis. When the number of predictors is large, as is usually the case for the data collected in surveys, the analysis of the joint effects of the predictors on the dependent becomes a serious practical problem due to the nature of the variables and the lack of a precise testable model.

Data description, data reduction, the identification of typologies, the identification of latent dimensions and the building of predictive models are all important aspects of survey analysis. All are particularly important when the purpose of the investigation is the generation of hypotheses and the estimation of the magnitudes of relationships rather than the testing of hypotheses formulated prior to the analysis. This volume of this book presents a number of analytical techniques which taken together should satisfy the needs of the practical survey analyst. The techniques are described briefly below.

Cluster Analysis

During the last decade the literature on techniques of numerical taxonomy or cluster analysis has grown at an enormous rate. Many new techniques have been developed which have been applied in fields as far apart as zoology and market research. The techniques are exploratory rather than analytical and are, in the main, hypothesis-generating rather than hypothesis confirming. The data for the techniques are in the form of observations on n individuals on p variables. These data are used to obtain a *matrix of similarities* or *dissimilarities* which represents the set of pairwise relationships between either the individuals or the variables. The similarity or dissimilarity coefficients can take many forms, including the product-moment correlation coefficient. However, the coefficients need not be parametric measures and are therefore particularly useful for the analysis of survey variables which are frequently not measured at the interval level of measurement. This matrix forms the basic input for the techniques, which can be classified as either *hierarchical* or *optimizing*. With hierarchical techniques an element, once classified in a cluster, remains in that cluster. Optimization techniques allow for the relocation of elements.

The judgement of the investigator plays a crucial rôle. The elements which are included in the analysis, the choice of similarity coefficient, the number of clusters formed are all reflected in the results. In many of the applications reported in the literature these problems are either ignored or sublimated. In Chapter 3 the methods of cluster analysis are reviewed, with special emphasis being given to the problems associated with using the methods in practice, and to the evaluation of the results. A particular problem which is common to many multivariate techniques is considered: whether the method of analysis itself imposes a particular structure on the data or whether alternative solutions are permitted to emerge. Practical examples are used to illustrate the difficulties and the solutions where they exist.

Factor Analysis and Principal Component Analysis

Factor analysis is a generic term which comprises a set of widely differing techniques which have been developed since Spearman introduced the conceptual framework in 1904. *Principal component analysis*, on the other hand, is a well-defined procedure. Both are analyses of correlation/covariance which work within the framework of linear models. Basically they attempt to condense a set of variables into a smaller number of constructed, conceptual variables, that is, the factors or components, which have not been directly measured. Thus they combine aspects of data reduction with the identification of latent dimensions. The observed relationships between the observed variables (the manifest space) are seen to be due to some underlying structure in the data which can be described in terms of the derived dimensions. Almost all factor and component methods have two quite distinct stages: first, the identification of the approximating space, and second, the specification of the coordinate axes within the space by which to describe it.

Chapter 4 presents a review and evaluation of the methods of principal component and factor analysis. Principal component analysis is described both in terms of its use in the simplification of the structure of multidimensional relationships and the use of component profiles as a breakdown of the respondent's profile. The problems of scaling, that is, of deciding between correlation coefficients and covariances as input, are discussed. Two aspects of factor analysis are described: the use of factor analysis as a model-fitting technique and its use in providing structure simplification amongst multivariate correlations. Factor and principal component analysis are compared and contrasted and particular attention is given to the vital and frequently misunderstood problems of axis rotation.

Latent Structure Models

In its strictest sense a latent structure model should be seen as a probabilistic scaling model but it can be, and often is, used as an exploratory or data reduction technique. Once the structural parameters have been estimated they provide information on the underlying dimensions as well as providing a useful way of contrasting and differentiating between the elements being studied. In common with other multivariate methods latent structure models aim at abstracting from the observed data a hidden pattern or structure based on the latent dimensions. A particular model is assumed in this process and this is described in Chapter 5 together with some practical applications. The basic axiom of local independence and the use of accounting equations are discussed and different forms of latent structure models are presented. The methods of estimation and the difficulties which arise in dealing with sample data are illustrated. Some guidelines are given for interpreting the results and for fitting and testing various models.

Multidimensional Scaling

Having carried out a survey of individuals or groups an investigator may wish to describe the 'typical' or 'average' individual, group, dwelling or firm. When the analysis deals with only one variable this description may easily be achieved by presenting the mean or some other measure of location together with information on the dispersion about that location. When many variables are involved this may still be done but the variables may be presented one at a time. The main interest of a social researcher will however be in patterns of variation and multivariate relationships. The problem is often one of describing how individuals are similar to one another and also about how they differ from one another in terms of sets of variables taken together. Chapter 6 describes a class of procedures known as multidimensional scaling [MDS] which are an important set of basically descriptive techniques which can help to solve this problem. The major advantage of the techniques described is that they are non-metric; they do not make assumptions about the level of measurement of the variables being analysed. This is important since 'the assumptions which are made about the levels of measurement of one's data and those contained in the scaling models used to analyse them are not merely convenient fictions, but commit one to substantive hypotheses about human behaviour' (Coombs, 1964). The freedom from metric assumptions means also that the same basic procedure can readily be extended to very different types of data. In Chapter 6 the basic non-metric MDS model and its extensions are presented and discussed and a detailed example of the application of the individual differences scaling model is given.

Asymmetric Analysis

The methods discussed above all deal with the analysis of a symmetric set of variables in the sense that all the variables are treated alike in the analysis. Frequently in survey analysis one or more of the variables will have a special significance. The purpose of the analysis may be the explanation of the variation between the scores of the individuals on these variables. For example a study of absenteeism might collect data on a number of indexes of absenteeism and also collect observations on a large number of additional variables. The purpose of the analysis might be to construct a predictive or explanatory model of the way in which the additional variables could be used to predict or explain the variations in absence behaviour between individuals. In this sense the analysis is *asymmetric*. The full set of variables (15) is included in the analysis. Many problems arise in such analysis using survey data of which the most important may be the following: (i) The explanatory or predictor variables (the x's) may be at different levels of measurement. Some may be classifications e.g. sex, race, marital status. More may be ordinal, i.e. occupational status, attitudes, etc. And some may be measured at interval level: age, income, etc.; (ii) The predictor variables may be intercorrelated. This problem is made more difficult to handle by the fact that many of the predictors may be classifications and the problems of assessing intercorrelations between such variables is more difficult than for continuous variables; (iii) there may be interaction effects present. The effect on the dependent variable(s) of changes in score or position on one of the independent variables may differ according to the individual's position on the other independent variables. In other words, the effects of the predictor variables are not additive. There are a number of reasons why such interactions may arise in the model. First, the variables observed may only be proxies for the underlying constructs, or each may measure some aspects of more than one construct, in which case their effects can be expected to be irregular. Second, an underlying theoretical construct may require several variables in order to measure it. An excellent example is provided in Kish and

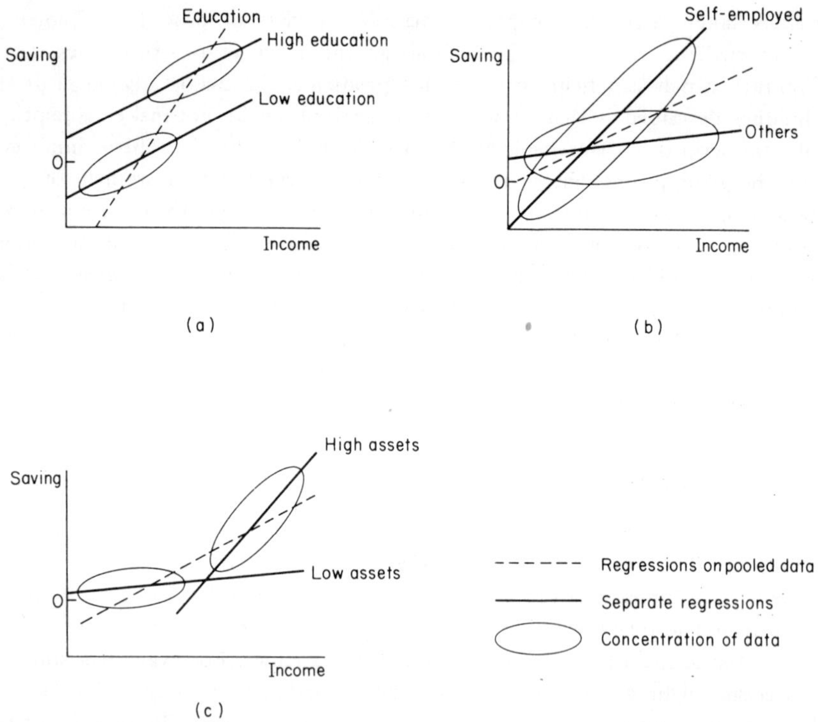

Figure 1.1. Intercorrelation *vs.* interaction:

(a) Multicollinearity. Additive model would represent both effects adequately.
(b) Interaction. Additive model would show no effect of self employment.
(c) Interaction and multicollinearity. Additive model would produce single income effect and single asset effect – not correct.

(Reproduced with permission from J. N. Morgan and J. A. Sonquist, *Journal of American Statistical Association,* **58**, 415–434 (1963))

Lansing (1957) where the 'family life cycle' concept is a complex combination of age, sex and number and age of children.

The problems of intercorrelation and interaction are frequently confused. An example is presented in Morgan and Sonquist (1963) which illustrates the differences between them and the effect when both are present for the case of a single dependent variable [see figure 1.1].

The functional form of the asymmetric relationship being examined can be represented in general as

$$(y_1, y_2, \ldots, y_m) = f(x_1, x_2, \ldots, x_p). \tag{16}$$

This functional relationship will not however be exact and a disturbance or error term should be included. This may be justified on any or all of three grounds. First, not all possible predictors will ever in practice be observed. Second, the randomness of human behaviour will introduce an unpredictable variation into the observed measures. Third, there will probably be measurment errors in the observation of the variables. The model should therefore be written as

$$(y_1, y_2, \ldots, y_m) = f(x_1, x_2, \ldots, x_p) + \epsilon \tag{17}$$

The state of knowledge in most fields of study in which survey research methods are used is such that the precise form of the relationship will not typically be stipulated in advance. Consequently the use of inductive procedures promises greater gains than the formulation of trivial testable hypotheses. A systematic 'search' of the data in order to identify a model which maximizes the explanatory power of the data is indicated.

The Automatic Interaction Detection technique (Sonquist and Morgan, 1964) is probably the most widely used such inductive method. It is discussed and evaluated in Chapter 8 in the context of a class of such procedures. The basic algorithms and some of their modifications are described. Some guidelines are given for the use and interpretation of the results, and the rôle of the method in hypothesis generation and model building is described.

The various methods described for symmetric analysis are in a sense in competition with one another. Each can provide a means of data reduction and summary, all can be used to some extent for constructing typologies and for the identification of latent dimensions. The crucial differences depend on the assumptions underlying the models and the degree to which a structure is imposed on the data. In each of the chapters in this volume the comparative strengths and weaknesses of each method in relation to its competitors are evaluated.

Chapter 7 provides another approach to the particular and important problem of index construction in attitude measurement in the context of questionnaire design. Both factor analysis and cluster analysis are applied to aid in the development of the attitudinal index and the relationship of the approach adopted is compared to the other available techniques.

All of these techniques can also be viewed as a preparatory stage for subsequent asymmetric analysis. The reduced data set or the scores on the latent dimensions can be used as part of a predictive or explanatory model. A similar analysis can be carried out on the set of dependent variables before the asymmetric analysis is carried out. It should always be stressed that all of this analysis is inductive and is appropriate at a particular stage of scientific inquiry. The aim is to gain insight into the structure and nature of relationships existing between variables. Where the form of the relationships is not specified *a priori* the models are inherently untestable. For some of the techniques some theoretical foundations exist which enable some tests to be carried out. But in all cases the effect of the sample design, particularly for complex sample designs, is not known and the interpretation of the analysis should therefore be extremely cautious. The techniques should be seen as powerful but dangerous instruments where the rewards for judicious use are high but the penalties for careless application are even higher. The basic principle is hypothesis generation and not confirmation. The investigator should not allow himself to be carried away by the apparent elegance of the results. The main purpose should be the estimation of the magnitude and direction of relationships in order to facilitate the formulation of specific and testable hypotheses or models for further research.

1.5.3 Model Fitting

The state of knowledge about the subject area may be such that the precise functional form of the relationship between the observed survey variables

$$(y_1, y_2, \ldots, y_m, x_1, x_2, \ldots, x_p) \tag{15}$$

may be specified in advance. Thus a model of the form

$$(y_1, y_2, \ldots, y_m) = f(x_1, x_2, \ldots, x_p) + \epsilon \tag{17}$$

can be stipulated where the functional relationship f is known. This enables the researcher not only to fit a model but also to test the goodness of fit of the model to the observed data. The specification of such a model implies far more prior knowledge than exists in the situations discussed in Section 1.5.2. The substantive theory must be considerably more exact in order to specify not merely the variables of interest but also the relationships between them. The main advantage of specifying the model *a priori* is that it is possible to test the fit of the model to the data. The main disadvantage is that specifying and fitting an inappropriate model may mask or obscure important relationships of other kinds in the data. It is important to balance the one against the other. A model is at best an approximation to the real world. The basic features of a good model should be that it summarizes or describes a complex situation by sacrificing some detail but that it preserves the essential elements of the structure. The problem in practice is complicated by the fact that the data are observed for only a sample of the population. Furthermore it is unlikely that *all* variables relevant to the relationship have been observed and the observation and measurement processes are probably subject to error. The disturbance or error term ϵ in (17) is therefore crucial in terms of the interpretation of the results. The model cannot be expected to fit the data exactly.

The class of models most widely used in survey analysis is the class of *linear models*. Each dependent variable is assumed to be a linear function of the parameters of the model. In other words,

$$f(y) = \sum_{j=1}^{p} \beta_j x_j + \epsilon \tag{18}$$

where $f(y)$ is a function of the dependent variables and $\{\beta_j\}$ are the parameters and the $\{x_j\}$ are the explanatory or predictor variables. Interactions can be built into the model by defining new variables which represent the interactions. For example, if the model for the set of variables (y, x_1, x_2) is

$$y = x_1 + x_2 + x_1 x_2 + \epsilon \tag{19}$$

by defining $x_3 = x_1 x_2$, we can write

$$y = x_1 + x_3 + x_2 + \epsilon \tag{20}$$

which is linear in x_1, x_2 and x_3. Although in principle many complex relationships can thus be represented as linear models, it is necessary to specify in advance the terms and variables in the model. With many explanatory variables, it is normally not desirable to include in the model all possible interactions between them. But the choice of variables in the model is crucial in determining its appropriateness to the data. This is the problem which led to the development of the search techniques mentioned in the previous section. It is sometimes necessary to carry out some preliminary analysis on the data in order to ascertain which terms should be included. But using the same data set to discover and subsequently to fit a model invalidates the criteria for testing the model.

The General Linear model

The *general linear model* (GLM) includes all models of the form (18) regardless of the assumptions made about the disturbance terms ϵ. Chapter 2 of Volume 2 describes the GLM and discusses in some detail the problems associated with its use. The most commonly used form of the GLM is *ordinary least squares regression*. In addition to assuming the appropriateness of a linear model, additional assumptions must be made about the distribution of the ϵ's if the parameters are to be estimated optimally. The disturbances must be uncorrelated with each other and with the x's and must all have equal variance. The *analysis of variance* is a special case of multiple regression which arises when all the explanatory variables are nominal-scale or categorical variables, while the dependent variable is still a continuous interval-scale variable. The *analysis of covariance* may be used when some of the explanatory variables are continuous and some are nominal. The continuous variables are used largely as a means of controlling what would otherwise be confounding variables in the interpretation.

The assumptions about the disturbances which are required in ordinary least squares regression are frequently not appropriate in practice. In some cases, however, techniques are available which permit best linear unbiased estimation of the parameters even when the assumptions are not satisfied. Two such techniques, *generalized least squares* and *instrumental variable estimation* are described in Chapter 2 of Volume 2. The latter technique is designed to deal with problems of measurement error in the explanatory variables, which occur frequently in survey work, particularly in the form of response errors which are discussed in Chapter 7 of Volume 2.

The procedures mentioned above are all restricted to the case of a single dependent variable. In many situations there will be several dependent variables which should be considered simultaneously. Two ways of dealing with such situations are *multivariate regression* and *simultaneous equation estimation.* The difference between multiple regression and multivariate regression is that in the latter case there will be more than one dependent variable. For simultaneous equation estimation a variable may appear as dependent in one equation and as an explanatory variable in others. In *canonical correlation analysis* the aim is to find the linear combination of the $\{x_j\}$ which has maximum correlation with a linear combination of the $\{y_i\}$. The first canonical correlation coefficient is the correlation between the 'best predictor' based on the explanatory variables and the 'most predictable criterion' based on the dependent variables. *Discriminant analysis,* on the other hand, is concerned with obtaining a linear function of the observations which permits the maximal separation of two or more groups of *individuals*.

Two particular cases of the general linear model receive special attention in Volume 2. The theoretical framework for each is presented in Chapter 2 of Volume 2. The first is *path analysis* which has become one of the standard tools for analysing interval-level data for which an explicit causal structure can be stipulated. The treatment of path analysis in Chapter 13 is based on the regression approach. This has the advantage of permitting a comprehensive coverage of model building, in particular, strategies for imposing constraints on the model. The testing of models with constraints against the unconstrained model is also discussed. The initial assumption in path analysis is that there is an explicit predictive ordering between the variables of the model. Inspection of the data will not produce such an ordering and it must therefore be specified *a priori*. The selection of the model is crucial since the data will always fit the model. This is not peculiar to path analysis since by sufficiently strong assumptions we can always force the data to conform simply by classifying departures from the model as 'error' or 'disturbance'. It does mean, however, that the contribution of the substantive theory is all-important. Chapter 3 of Volume 2 provides a practical guide to path analysis and discusses the use of residuals in exploring the data. It should always be borne in mind, however, that the choice of a particular model or set of models concentrates attention on particular aspects of the data and may obscure other equally, or perhaps more, important relationships.

The *log-linear model* is a special case of the general linear model. It is a method of ascribing a functional form to the relationships between a set of categorical variables $\{y, x_1, x_2, \ldots, x_p\}$. There are two principal cases, *symmetric analysis* in which all of the variables are given equal status in the analysis (there is no y-variable) and *asymmetric analysis* where one (or more) of the variables is dependent and the orientation of the analysis is towards explaining or predicting this variable by means of combinations of the others. The analysis is concerned with the explanation of the pattern of frequencies in the cells of a multiway table in which each variable (factor) is represented by a single dimension. Each cell in the table is defined by the factor combination it represents. In the analysis of variance model (see Chapter 2 of Volume 2) the analysis concentrates on explaining the variation between mean scores on a dependent variable in terms of the effect of each factor considered separately and the joint effects of the factors. The log-linear model, which is discussed in Chapter 4 of Volume 2, is similar in that the explanation is in terms of factor levels and factor combinations but differs in that the focus of the analysis is on the distribution of the elements between cells and not on means or scores within these cells. For many survey variables the level of measurement is either nominal or ordinal and treating the numerical assignment of scores to the classes as interval-level data is inappropriate. The analysis particularly for nominal-scale data must therefore concentrate on the frequencies rather than on any other measures derived from the observations. The loglinear model provides a framework within which such analysis is possible.

Analysis over time

Classical sampling theory for finite populations is concerned with making inferences about variables x_1, x_2, . . . at a particular point in time. It is thus concerned with providing a static representation of a population. It is however possible to consider the population as a dynamic system which changes over time. For this purpose the simple picture of a fixed population must be extended in two ways. First, the variables of interest must be thought of as changing in time, probably in a random manner. In other words x_1, x_2, . . . now become realizations of stochastic processes. Second, the composition of the population may change as new individuals enter and old ones leave. Inference in such situations can thus centre either on the individual stochastic processes or on the changing aggregate structure of the population. Chapter 5 of Volume 2 describes some methods of analysing survey data designed to yield information about the population viewed as a dynamic system.

If the interest is simply in charting the changing character of the population it will be sufficient to sample in the ordinary way at intervals of time. In this way it will be possible to construct time series of estimates of population parameters and their relationships. More often it will be necessary to delve more deeply in order to study the process of change itself whether the purpose is exploration or prediction. This will require the estimation of quantities such as rates of change as well as means and proportions. It should be emphasized that adequate analysis of data from dynamic systems requires an explicit (rather than an implicit) model of the underlying process. This, in turn, has implications in terms of the kind of data required.

Many of the problems in this area resolve themselves into one or the other of the following forms:

(1) Members of the population can be classified at any point in time as being in one, and only one, of k mutually exclusive categories. The numbers in these categories change in a random way as a result of the operation of probabilities (or rates) of movement between categories. Data on numbers in the categories and the flows between them can be used to estimate the parameters of a flow model and test its goodness of fit.

(2) Members of the population can be characterized by the (random) length of time they spend in a particular category. It is therefore necessary to devise methods of estimating such distributions and this can be done using either cohort or census data.

The discussion in Chapter 5 of Volume 2 is illustrated by reference to data on residential, occupational and social mobility.

1.6 CONCLUSION

Survey research may be viewed as the systematic collection of information from a sample in order to describe, explain, understand or predict some characteristics of the population. The usefulness of the information collected is dependent jointly on all the stages of the survey and this unity must be preserved in the analysis. The set of data with which the survey analyst is faced has little or no meaning except in the context of the processes which generated it.

Following Popper (1963), 'What we should do is to give up the idea of ultimate sources of knowledge, and admit that all knowledge is human; that it is mixed with our errors, our predjudices, our dreams and our hopes; that all we can do is grope for truth even though it is beyond our reach.' What is necessary is to systematize our search for knowledge by recognizing explicitly the sources of error of which we are aware. In survey research the most fundamental constraint is that imposed by the nature of the survey variables and the level at which they are measured. In addition there are three sources of error which can, and should, be treated as an inherent part of the analysis: nonresponse, sampling errors and response errors. The technical sophistication of our analytical methods should not be permitted to conceal from us the weaknesses of the data on which we mean to base our conclusions. Neither should we use a complex technique if the answer to the substantive question we wish to ask can be obtained by using a simple one. The justification for complicating the analysis must be that it is *necessary* to do so in order to simplify or clarify our knowledge of the subject area.

REFERENCES

Blalock, H. M. and Blalock, A. B., eds. (1968), *Methodology in Social Research*, New York: McGraw-Hill.

Cochran, W. G. (1953 and 1963), *Sampling Techniques*, New York: Wiley.

Coombs, C. (1953), 'Theory and methods of social measurement', in L. Festinger and D. Katz (eds.) *Research Methods in the Behavioral Sciences*, New York: Holt.

Coombs, C., Dawes, R., and Tversky, A. (1970), *Mathematical Psychology*, New Jersey: Prentice-Hall.

Deming, W. E. (1950), *Some Theory of Sampling*, New York: Wiley.

Erdös, P., and Renyi, A. (1959), 'On the central limit theorem for samples from a finite population', *Pub. Math. Inst.*, Hungarian Academy of Sciences, *4*, 49–57.

Hájek, J. (1960), 'Some contributions to the theory of probability sampling', *Bull. I.S.I.*, *36/3*, 127–134.

Hansen, M. H., Hurwitz, W. N., and Madow, W. G. (1953), *Sample Survey Methods and Theory*, New York: Wiley, Vols. I and II.

Hollander, M. and Wolfe, D. A. (1973), *Non-Parametric Statistical Methods*, New York: John Wiley and Sons.

Kalton, G., and Blunden, R., (1975), 'Sampling Errors in the British General Household Survey', (to be published).

Kish, L., (1959), 'Some statistical problems in research design', *American Sociological Review*, **24**, 328–338.

Kish, L., (1961), 'Efficient allocation of a multi-purpose sample', *Econometrica*, *29*, 363–385.

Kish, L., (1965), *Survey Sampling*, New York: Wiley.

Kish, L., (1971), 'Multipurpose programs for sampling errors', *Proc. of I.S.I.*, II, 216–221.

Kish, L., and Frankel, M. R. (1974), 'Inference from complex samples', *Journal of the Royal Statistical Society*, *Series B*, **36**, 1–37.

Kish, L., Frankel, M. R., and Van Eck, N. (1972), *Sampling Error Program Package*, Institute for Social Research, Ann Arbor.

Kish, L., Groves, R., and Krotki, K. (1975), 'Sampling errors for fertility surveys', (to be published).

Kish, L., and Lansing, J. (1957), 'The family life cycle as an independent variable', *American Sociological Review*, **22**, 512–519.

McGinnis, R., (1958), 'Randomization and inference in sociological research', *American Sociological Review*, **23**, 408–414.

Morgan, J. N., and Sonquist, J. A. (1963), 'Problems in the analysis of survey data: and a proposal', *JASA*, **58**, 415–434.

Moser, C. A., and Kalton, G. (1971), *Survey Methods in Social Investigation*, Heinemann, London.

Nathan, G. (1972), 'On asymptotic power of tests for independence in contingency tables from complex stratified samples', *JASA*, **67**, 917–920.

O'Muircheartaigh, C. A., and Wiggins, R. D. (1976) 'Sample design and evaluation for an occupational mobility study', *Economic and Social Review*, **8**.

Scott, C. (1961), 'Research on mail surveys', *JRSS(A)*, **124**, 143–205.

Selvin, H. (1957), 'A critique of tests of significance in survey research', *American Sociological Review*, **22**, 519–527.

Sonquist, J. A., and Morgan, J. N. (1974), *The Detection of Interaction Effects*, Ann Arbor: Institute for Social Research.

Sukhatme, P. V. (1954), *Sampling Theory of Surveys with Applications*, Ames, Iowa: Iowa State College Press.

United Nations (1950), *The Preparation of Sampling Survey Reports*, New York: U.N. Series C, No. 1. Also (1964), Series C, No. 1, Rev. 2.

Yates, F. (1949), *Sampling Methods for Censuses and Surveys*, London: Griffin.

Chapter 2

The Preparation and Processing of Survey Data

C. Payne

2.1 INTRODUCTION

Nowadays computers play a fundamental part in the analysis of survey data and most of the techniques described in this book are unusable without them. Despite this, the published work on the practicalities of computer analysis is almost entirely confined to manuals for particular packages. This lack is due to the wide variations in computer hardware and programs that exist between installations, so that the particular knowledge required depends largely on the local situation. However there are some good general principles about the handling of survey data which minimise the difficulties that arise.

There are a great many computer programs for survey analysis. Anderson and Coover (1972) review the main American ones, and the SSRC Newsletter (1973, 1974, 1975) brings out occasional supplements on British programs. The programs vary widely in function, availability and ease of use, but they can be grouped into a few main types. First are the programs whose main function is to produce descriptive statistics in a compact and readable form; these are typically found in market research organizations. A second group consists of the large social science packages used in most universities and research institutions. These have comprehensive facilities for basic data management and description, plus a limited range of analysis techniques. Most readers of this book will find themselves using such packages at one time or another, and details of some of the more important ones are given in the Appendix to the chapter. There are also a large number of programs covering various aspects of survey processing available at particular installations. Thirdly there are programs for particular statistical techniques; most of the programs described in the other chapters fall into this category.

The types of program described above are usually not sufficient for survey processing. Utility programs specific to each computer installation are needed for basic operations such as copying, listing, sorting and editing files. In addition the user has sometimes to provide his own special purpose programs to deal with non-standard features of the data file or particular requirements in analysis. These are written in a high level computer language (usually FORTRAN although there are other possibilities), but the computer's machine language may be necessary for particularly difficult problems.

2.2 THE SURVEY DATA FILE

2.2.1 The Data Matrix

The starting point for all the analysis techniques described in this book is the rectangular data matrix giving the values of the variables for each case. Some techniques are based on a summary of the variables in this matrix, for example the correlation matrix (factor analysis), others on a summary of the relationships between the cases such as a similarity matrix (cluster analysis). In one-off surveys where the same questions are asked of each respondent the data matrix is obtained straightforwardly. However, some surveys have a more complex data structure which must be reduced to a rectangular data matrix before analysis can proceed, and this may result in many cells of the data matrix being empty.

One complex type is the *hierarchical* structure where there are several levels of interest. Figure 2.1 gives an illustration; the sampling unit is the dwelling, within which data is collected on households and within households on individuals of different types.

The cases can be defined in various ways here such as dwellings, adult males or children. The variables required for cases at a particular level may include summary statistics of variables at a lower level (e.g. mean income of individuals for households), at the same level (e.g. number of children for fathers) or at a higher level (e.g. number of bedrooms in household for children). A problem which arises frequently with hierarchical structures is that some levels are missing in a particular sampling unit (e.g. there is no adult female for Dwelling 2 in Figure 2.1) with the result that parts of the data matrix are empty.

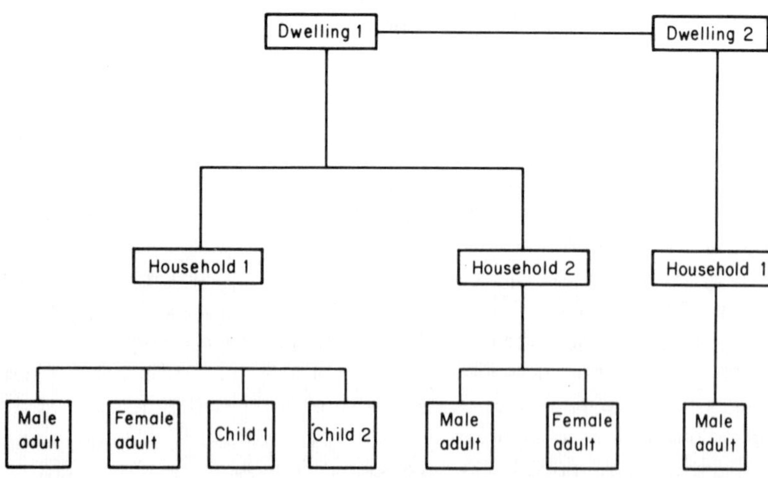

Figure 2.1. A hierarchical structure.

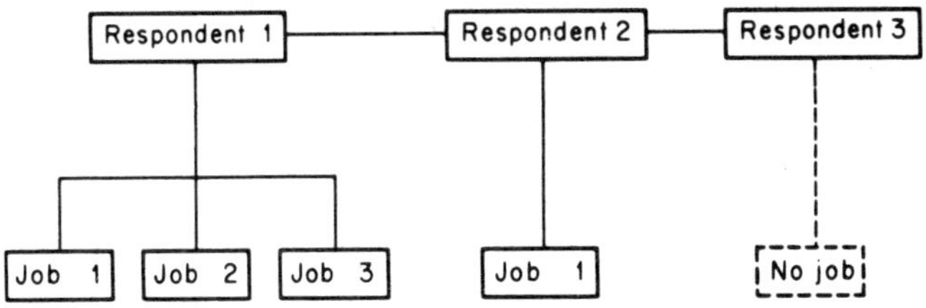

Figure 2.2. A tree structure.

Another complex type of structure is the *tree structure* where some variables apply only to certain cases although all variables are at the same level. Figure 2.2 gives an example where the sampling units are adult males and information is collected on each of the jobs they have had.

Typically this structure contains a 'root' segment of variables which apply to all cases and 'branch' segments containing variables applicable only to some cases. There are two ways of 'rectangularizing' this structure depending on the type of analysis required. One method would take respondents as the cases with variables such as first job, second job, current job and age and the resulting matrix could have much missing data. A second approach would take jobs as the cases, e.g. first job or job at 45.

A panel survey where a sample of respondents is contacted at successive times ('waves') gives rise to the particular type of tree structure illustrated in Figure 2.3. This structure is fairly straightforward to adapt to the rectangular form.

Its main feature is that respondents tend to drop out of the survey or cannot be contacted at particular times (new respondents can be introduced at later stages). The cases for such structures may be defined in several ways, e.g. all respondents interviewed at times 2 and 3 or all respondents ever interviewed. If in the latter case the variables are defined over the whole time period there will generally be much missing data.

Cohort studies present special problems in the construction of variables if the intervals at which the data are collected are not the same for each respondent. Adjustments may have to be made to the measurements so that values are obtained for particular target dates or rates of change calculated for time periods of fixed lengths.

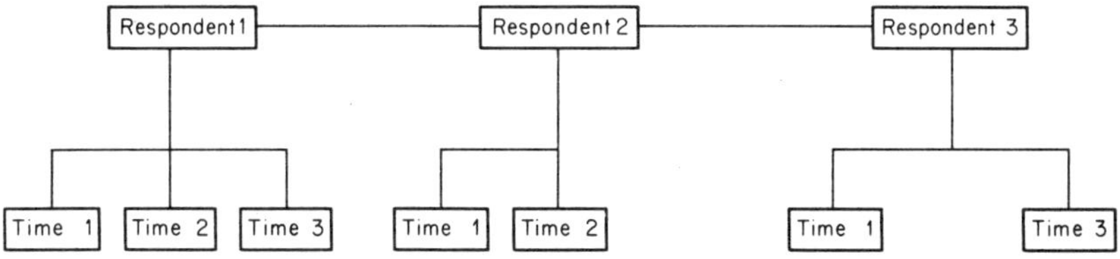

Figure 2.3. A tree structure for a panel survey.

Some surveys give rise to complex structures which combine *both* the hierarchical and the tree types (e.g. a repeat survey of households) but the basic principle of reduction to a rectangular matrix still holds. Very few packages can handle complex structures and special programming is usually necessary to abstract the matrices required.

2.2.2 Special features of survey files

Data matrices derived from social surveys have certain features which dictate how they are handled in both file construction and analysis.

(a) *Size*. Survey data files are often very large with matrices of more than 1000 cases and 100 variables being common. This presents certain problems in the organization and processing of the data; for example, the variables for each case may be spread over a number of punched cards which have to be ordered correctly in the file. Because of limitations in computer store and time, analyses often have to be restricted to subsets of variables or cases so that the computer system must include facilities for creating work files from the basic data file.

(b) *Subgroups*. Frequently the cases in the data matrix are partitioned according to some meaningful division in the data such as region and analyses are required both on all the cases and separately for each category or combination of categories for the analysis of selected cases with special characteristics.

(c) *Derived Variables*. Often the original data matrix needs to be extended either by the addition of new variables derived from the existing ones (e.g. age from birth date) or by merging the data file with other files (e.g. waves from a panel survey).

(d) *Differential Coverage*. The results must be adjusted to allow for any differential coverage of the target population. This problem arises when there are known variations in non-response (for example in a panel survey where certain types of respondent drop out more quickly than others), when a sample allocation procedure has been used which is disproportionate for different sampling units or when post stratification is required (see Chapter 1).

(e) *Sample Design*. Many surveys use a complex sample design which must be taken into account when calculating sampling errors for the statistics of interest.

(f) *Missing Data*. Invariably social survey matrices have missing data, either because of refusals or because some questions do not apply to some respondents. As we have seen, the conversion of a hierarchical or tree data structure to a rectangular matrix also leads to empty cells. The existence of missing data has important implications for analysis since most analytic techniques require a complete data matrix. Various strategies exist for dealing with this problem such as the imputation of missing values or the omission of cases with missing values on the variables of interest.

(g) *Data Validation*. Social surveys are characterized by problems of measurement: interviewers ask the wrong questions or mark the wrong answer; respondents may misinterpret questions or give false replies; when completed schedules are returned, coders assign wrong codes and key punchers make mistakes. Even if there is a high degree of quality control over the data collection process errors still appear in the raw data. This means that an important part of the file creation process is the detection and correction of errors.

2.2.3 Stages in the processing of survey files

The special features described above lead to a succession of well defined stages in the processing and analysis of survey files. These are set out in Figure 2.4:

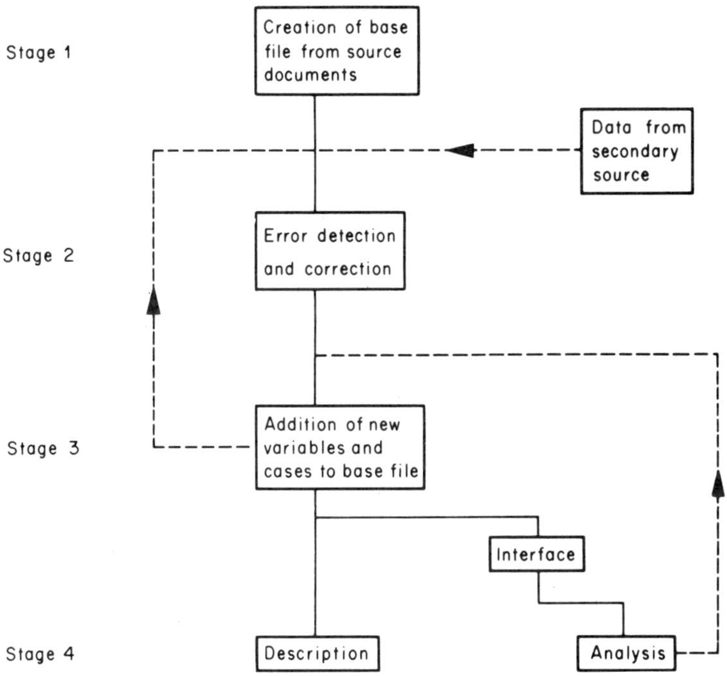

Figure 2.4. Stages in processing survey data.

Stage 1 covers the preparation of the base data file in machine readable form from the data source (questionnaire, tape recordings, etc.). At this stage the researcher must decide what information is to be recorded as well as the coding scheme and data formats.

Stage 2 deals with the detection and correction of errors in the base file, including those which arise in the process of assembling the file as well as those due to errors in the source document. If the survey data file is obtained from a secondary source such as a survey archive (usually on punched cards or magnetic tape) the file generation process will normally be entered at Stage 2.

Stage 3 is concerned with the extension of the base file to include derived variables, modified versions of the original variables, variables obtained from other sources and the addition of extra cases. Sometimes there is feedback to Stage 2, as, for instance, when waves of a panel survey are merged.

Stage 4 is the description and analysis of the data matrix. A pragmatic distinction between the two tasks is made in Figure 2.4, although some computer packages cover some aspects of both. Many of the computer programs available for the analysis techniques discussed in this book are either restricted in the amount and format of data they can take or can only handle a summary of part of the data matrix such as a contingency table. Thus there is a problem about their *interface* with programs that can process the basic data file. Other important considerations at this stage are the use of weights to allow for differential coverage, the calculation of sampling errors for the parameter estimates and the treatment of missing data in the statistical analysis. There may be feedback to the previous stage if variables derived from a statistical analysis such as factor scores are to be added to the permanent file.

2.3 STAGE 1: THE BASE FILE

2.3.1 Choice of Input Medium

Computer input is of two types. *Primary* media are those which the user can prepare himself and include punched cards and paper tape. *Secondary* media such as magnetic tape or disc require the intervention of a computer process but have the advantage of speedier processing and greater security. Although in most cases files on secondary medium must be constructed from a primary source, machinery for direct transfer to a secondary medium (e.g. key-to-disc system) is becoming more widely available.

There are several important reasons why punched cards are by far the most frequently used primary medium for survey data. Firstly card punching equipment is universally available and all computer packages can handle data on cards or in card-image form on a secondary medium. Secondly unit-record equipment such as card sorters or tabulators can be used for basic processing as an alternative to the computer (see Janda (1969)). Thus we shall concentrate here on the methods of recording data on punched cards.

2.3.2 Recording data on cards

In punching data on cards several basic principles should be followed.

(a) *Non-numeric codes should be avoided.* All computers have a basic set of about 64 characters which they can read, called the internal character set, and each character corresponds to a particular combination of holes in a column of a card. The set includes the numeric characters 0, 1 . . . 9. The remaining characters are called ‘alphanumeric’ and include the letters A . . . Z as well as special characters such as %, +, −. Most programs will read numeric codes, but many will not accept alphanumeric characters which in any case require conversion to numeric form for analysis to proceed. For this reason numeric codes should be used for the values of the variables; these can be either single or multiple digit integers.

(b) *Multipunched codes should be avoided.* These are combinations of holes in a card column which do not correspond to characters in the computer's internal set. In the past, many card files prepared for counter sorter analysis used multipunched codes for coding multiple response questions (for example, which newspapers are read) and for compressing several variables into one card column. However most packages will not accept multipunch codes and a laborious conversion to numeric form is required. A further problem is that computer hardware facilities for reading multipunched codes are not universally available. A multiple response variable should be coded as a sequence of binary variables. Moser and Kalton (1971, p. 421) have described an ingenious coding scheme for such variables called geometric coding. If there are R responses to a question then the single value coded is

$$\sum_{i=1}^{R} \alpha_i 2^{i-1}$$

where $\alpha_i = 1$ if the ith response is given, zero otherwise. This gives a compact solution but may be difficult to unscramble at the analysis stage.

(c) *A code should be assigned to every value of a variable.* In particular, missing values should be coded with different codes for refusal, not applicable and don't know. It is convenient for subsequent analysis if the same missing value codes are used for every variable whenever possible.

(d) *The variables must be punched in the same order for each case.* There are two ways of organising this. In *fixed-column* format the values of each of the variables are located in the same column or group of adjacent columns for every case. Where the case covers more than one card the 'same column(s)' means the same column(s) on the particular card in the case. Almost all programs will accept data in this format. Variables are read into the computer either by a Fortran-type format statement or by reference to the starting card/column number and width. In *free-field* format the variables are punched with some separator such as a blank or comma between them and do not necessarily occupy the same columns in different cases. Some programs will accept data in this format. It is best suited for small quantities of data as data checking can be a rather difficult process for most checking programs assume data in a fixed format. It is good practice to punch a variable to identify the end of each case; this helps considerably with checking.

(e) *Variables which can be derived from the basic data should not be punched.* Providing the programs used have good variable generation facilities (otherwise a special purpose program can be written) there is no point in punching such variables at this stage as this involves extra effort and is another source of error.

2.3.3 Information that should be recorded

Some additional information should be punched at this stage to identify cards and also to give sampling details for the case for use in subsequent analysis.

(a) *Identification*. Each card should contain a number uniquely identifying the case (usually a questionnaire number) and a number identifying each card within the case ('card-in-case' or 'deck' number). For practical convenience this information is usually punched at the beginning of the card. The information is required so that the file can be sorted into order and the source documents related to the cards. In some circumstances the identification number may usefully contain within it a code for some higher level of aggregation of the cases, for example region or sampling unit (see below). It is a good idea also to include a study number which is the same for every card in the file so that the file can be distinguished from other files.

(b) *Contact information*. Information on the particular interviewer, time, place and length of interview should also be recorded. This is required for the analysis of interviewer variance (see Chapter 7 of Volume 2) and of response errors, and also for field quality control in a sequence of surveys.

(c) *Sampling unit information*. In multi-stage surveys the sampling units for each stage should be recorded. In, for example, a survey with sampling stages constituency, ward, polling district and household, codes for the sampling units at each stage should be recorded. This information is used to examine differential non-response between stages of the survey and to calculate sampling errors taking account of the complex design. The sampling units codes can be put together to give a case identification number.

(d) *Aggregate level information*. In some surveys we may wish to include data from another source at a later stage. For example, in a survey of voting intentions constituency level variables such as percentages voting for each political party may be added. In order to do this a constituency code must be punched so that the survey file can be matched with the constituency file.

2.3.4 Card layout and organization

(a) Surveys with a simple structure. These have a fixed number of cards per case and the layout of the cards is straightforward as illustrated below.

Case ID ↓	Deck ID ↓		Study number ↓
001	1	Case 1 variables Deck 1	5032
001	2	Case 1 variables Deck 2	5032
002	1	Case 2 variables Deck 1	5032
002	2	Case 2 variables Deck 2	5032

The cases are arranged in sequence according to the identification number. In surveys where the cases are partitioned into groups for which separate analyses are required, the groups can be presented one after another (the identification number should then contain the group code). Some packages (for example, SPSS) can make explicit use of this ordering into groups. All cases contacted in the survey, regardless of whether complete information was obtained, should be included in the base file, so that analysis of non-response, response errors and interviewer variance can be made. Sampling units where complete refusals were obtained can be omitted at a later stage.

(b) *Complex structure*. In surveys with a complex structure additional complications are introduced because the data file usually involves a variable number of cards per case. The layout for *panel* surveys is fairly straightforward, extra cards being added to an already existing file as successive waves are carried out. Obviously the same identification numbers should be used for each case in each wave and incremental card-in-case numbers should be used so that the separate files can be matched and merged.

Hierarchical files are more complicated; the basic principle is to code at each level the identification number for each higher level. These can be organized either with all levels combined in one file or with a separate file for each level. With the first method the identification scheme is complicated since we make sure that the cards can be ordered properly. An example is given below.

Dwelling ID	Household ID	Person ID	Deck within level		
001	0	0	1	Dwelling 1 variables	
001	1	0	1	Household 1 variables	Deck 1
001	1	0	2	Household 1 variables	Deck 2
001	1	1	1	Person 1 variables	Deck 1
001	1	1	2	Person 1 variables	Deck 2
001	1	1	3	Person 1 variables	Deck 3
001	2	0	1	Household 2 variables	Deck 1
001	2	0	2	Household 2 variables	Deck 2

Note that the deck number within each level is in the same column for each level; this ensures that the file can be sorted on the identification field. If separate files are prepared, each file need only have the identification information for its own and higher levels.

A special purpose program is usually required to abstract matrices from a hierarchical file with a variable number of cards per case. There are several programming techniques which can be used to do this depending on what information about the structure of the file is coded. One technique requires information coded at the highest level about the number and types of cards found at each of the lower levels (such information is also useful for checking purposes). The program can then determine the complete structure of the cards for the sampling unit. An alternative technique takes one card at a time and examines the identification field to see which deck, level and sampling unit it refers to. With this technique information about the structure need not be coded, but there is the problem that we know that we have come to the end of the preceding level only when a new level is reached. In programming terms the technique either requires a card re-read facility (not generally available) or the card last read must be stored internally for the processing of the next level.

2.4 STAGE 2: DATA CHECKING

Checks must be made at several points during the assembly of a file. The logical sequence for these is shown in Figure 2.5.

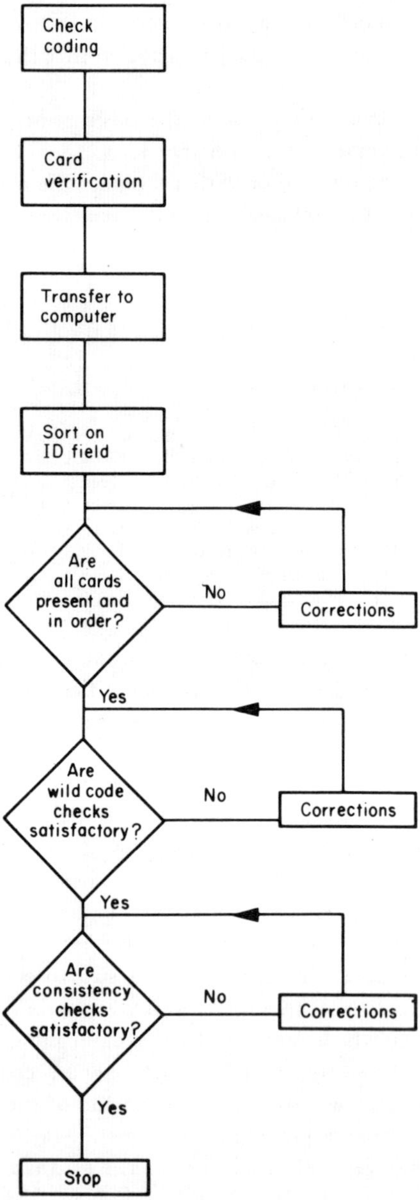

Figure 2.5. Stages in the data checking process.

Check Coding

In a carefully conducted survey the coding of information from questionnaires is checked as the coding is carried out, so that further errors can be avoided. Checking can be automated by having a second check coder code a sample of questionnaires, punching both the original and the check coding sheets onto cards, and comparing the two sets of cards by computer program.

Card Punching Verification

It is advisable to have card punching verified. The cards are repunched on a verification punch which stops when the operator attempts to punch a different character to that already on the card. This operation minimizes punching errors.

The cards can now be read into the computer and are usually copied to a secondary medium such as magnetic tape or disc. From this point on there will be frequent need for listings of the card file which should be numbered sequentially. Back-up copies of files should be kept in case of accidents.

Checking the Structure of the File

If necessary, the file is now sorted into order on the identification field. Most computers have utility programs for doing this, and some packages (e.g. OSIRIS) have their own sorting facility. The next stage is to check that all the cards are present and in the correct sequence. For files with a simple structure each case must have a fixed number of cards in card-in-case order. Errors in the identification number will lead to duplicate or missing cards. Sometimes utility programs are available for carrying out the sequence check and for correcting sequence errors by inserting or deleting cards. Where they are not provided the sequence check can be performed by visually scanning a file listing and resubmitting a manually corrected file.

The structure of complex files is more difficult to check by computer because the number of cards varies between sampling units. The only way to ensure that none are missing is to check the file listing at each level of every unit against the source documents.

Data Editing

Once the structure of the file is correct, we can look for errors in the data itself. The first step is to ensure that the values of each variable are permitted by the coding scheme: this is called a *wild code check*. Following this, *consistency checks* can be done for improbable or impossible combinations of values on two or more variables. It is also worthwhile to examine cases with extreme values on an interval level variable or on the ratio of two such variables. Even when these values prove to be accurate the information is useful in subsequent analysis, as cases with extreme values can bias parameter estimates.

Consistency checks are particularly important in surveys involving repeated measures of individuals over time. Here a wide range of checks can be made, for example on absolute and percentage changes and on the consistency of basic information such as age between each period. These issues are discussed by Goldstein (1970).

When errors have been located reference must be made to the source documents to determine the cause of the error and the correction required. If the correct value cannot be retrieved there are two alternatives: the value can be imputed or it can be coded as missing. Some imputation methods are discussed below.

The specification of consistency checks usually requires the flexibility of a high level language such as FORTRAN. For this reason few packages contain comprehensive checking facilities (OSIRIS being an honourable exception) and special purpose programs are needed. However packages which have good facilities for generating variables and which allow selective printing of variables can be used in the following way. Construct test variables which define the checks required and which take value 0 if a case passes the test, 1 if it fails, and then list the identification numbers and values of the test variables for each case in the file. Since most packages handle cases with a fixed number of cards only, this procedure cannot be used for files with a complex structure, where special purpose programs are required.

Editing is carried out either by replacing whole cards or card images in the file or by changing selected columns. On-line utility editing programs are especially suited for the correction of data files stored on magnetic disc.

Carrying out consistency checks is time consuming and expensive. The researcher should weigh the increased precision obtained against the cost of achieving this (see Richardson and Cunningham, 1972), bearing in mind that there are other errors, such as response errors, which may be much more important than those that can be detected by consistency checks.

Imputation of Missing Values

Missing data can either be handled by omitting cases with missing values during analysis in the way shown in Section 6 or by inserting imputed values at the data editing stage. Methods of imputation range from the purely *ad hoc* to ones involving strong distributional assumptions.

Interpolation of missing values is sensible in data involving repeated measurements over time. When this is not possible, the *hot deck* procedure is often used. A case with similar characteristics to the case whose missing value is to be imputed is selected at random, and its value on the variable in question is used instead. Clearly this method depends heavily on the definition of 'similar characteristics'.

With interval level data missing values are sometimes replaced by mean values although this distorts statistics such as variance. An alternative approach is to use a regression model to predict the variable whose values are missing, including in the calculation all cases for which there is complete information on both the dependent and independent variables. The coefficients obtained can then be used to substitute for the missing values. This procedure can be very complicated because different cases usually have different patterns of missing values and some averaging over regressions is required. Elasoff and Afifi (1966) describe extensions of the regression procedure which are appropriate when the missing values can be assumed to occur randomly.

Hocking *et al.* (1974) give a method which assumes that the data follows a multivariate normal distribution and uses a maximum likelihood estimation procedure. Each case is regarded as an observation $(X_1, X_2 \ldots X_p)$ on the p-variate normal population $N(\boldsymbol{\mu}\ \boldsymbol{\Sigma})$ and $\boldsymbol{\mu}$ the vector of means and $\boldsymbol{\Sigma}$ the variance-covariance matrix are estimated from both the complete and incomplete data. Transformations of variables to achieve multivariate normality are suggested. Values are imputed using regression equations involving the estimates $\hat{\boldsymbol{\mu}}$ and $\hat{\boldsymbol{\Sigma}}$. If, for example, X_1 is known but X_2 and X_3 are not then the values imputed are

$$X_2 = \hat{\mu}_2 + \frac{\hat{\sigma}_{12}}{\hat{\sigma}_{11}}(X_1 - \hat{\mu}_1) \quad \text{and} \quad X_3 = \hat{\mu}_3 + \frac{\hat{\sigma}_{13}}{\hat{\sigma}_{11}}(X_1 - \hat{\mu}_1)$$

where $\hat{\sigma}_{ij}$ are the elements of $\hat{\boldsymbol{\Sigma}}$ for the covariance of X_i with X_j.

Freund and Hartley (1967) propose a least squares correction procedure which can be used where the relationships between variables can be specified by one or more consistency equations (for example, that the sum of variables must equal a check total). If x_j and y_j are the corrected and uncorrected values for a variable in the jth case, their procedure chooses x_j to minimize the sum of squares of the discrepancies in the consistency equations:

$$\sum_j^n w_j(x_j - y_j)^2 + \sum_i^m u_i \left(\sum_j^n a_{ij}x_j\right)^2 \qquad \begin{array}{l} j = 1, \ldots n \\ i = 1, \ldots m \end{array}$$

where w_j is a weight given to the jth case (so that 'good' data can be used to estimate 'bad' data),

$$\sum_j^n a_{ij}x_j$$

is the ith consistency equation with the constraint

$$\sum_j^n a_{ij}x_j = 0$$

and u_i is the weight given to the ith consistency equation (some are more important than others). The solution to the minimization, in matrix terms, is

$$\mathbf{x} = \mathbf{D}_w + (\mathbf{A}'\mathbf{D}_u\mathbf{A})^{-1}\mathbf{D}_w\mathbf{y}$$

where $\mathbf{D}_w$ is the n x n diagonal matrix of weights for each case
$\mathbf{y}$ is the original data (n x 1)
$\mathbf{x}$ is the corrected data (n x 1)
$\mathbf{A}$ is the matrix of coefficients for the restrictions a_{ij}
and $\mathbf{D}_u$ is the diagonal matrix of weights from the consistency equations.

This procedure can be easily programmed so that the data editing can be done automatically. The authors suggest that it should be used after wild code checks and consistency checks have been done and that the number of corrections made for each case in this preliminary stage could be used in the computation of the case weights w_j. The procedure would seem to have limited application to surveys of human populations where consistency equations are difficult to define but is useful in surveys where the data collected is mainly interval level, for example in studies of firms.

Although it is clear that the choice of imputation method must depend on the particular characteristics of the data, there is little published work on the circumstances in which they are appropriate or on the relative advantages of each. No matter which method is used, one must beware of building into the data a tendency automatically to confirm the hypotheses they are supposed to test.

2.5 STAGE 3: PREPARING FILES FOR ANALYSIS

When data checking is completed, the file has to be got into final form for analysis. There may be several jobs to do at this point.

(a) *Transformations*

Transformations of variables are often needed in analysis and the transformed variable either replaces the original or is added to the file as a new variable. Some common transformations on single variables include converting a continuous variable into a categorical one, combining codes of categorical variables, creating dummy variables, standardizing, and changing the shape of a distribution by, for example, a log or square root transformation. New variables are also defined as a function of two or more variables, or as the sum of values on a set of variables, and sometimes they are derived from an analysis of the data, as when factor scores for each case are obtained from a factor analysis.

Transformations require the flexibility of a high level language, and most analysis packages have some provision. Some, like GSP, allow the user to insert his own Fortran routines, while in others (SPSS, BMD) he gives a Fortran-like specification. However there are some facilities which are especially useful but which are not universally available. One such concerns the treatment of missing values in transformations: OSIRIS and SPSS cater for these automatically by setting the result variable to a missing value code if one or more of the values on the original variables is missing, thus avoiding the need to test each variable for missing values. Another involves those transformations like factor scores which depend on the results of a statistical analysis. The transformed values are generated during the course of the analysis, but few packages allow the user to retain them automatically as a new variable in the file. Without this facility, the transformed variables must be added to the file in a separate run. OSIRIS's *tabular recode* facility is valuable for transformations which assign values to the derived variable according to the combination of values on two or more categorical variables (as when occupational prestige is defined by type of occupation combined with the position held in the profession). Performing such transformations is very tedious if each combination has to be specified separately, but the tabular recode allows the user to input a table of new values with two or three defining variables.

(b) Weighting

When a sample has differential coverage of the target population (see Section 2) each case should have a weight associated with it which is used in calculating estimates of population parameters. If, for example, having stratified by sex, a sampling fraction of 1/100 is used for males and 1/150 for females, each male should be given a weight of 1.0 and each female a weight of 1.5. In practical terms this means that a variable is added to the file giving the weight to be applied to cases with particular combinations of characteristics. The use of weights in analysis is described in Section 6.

(c) Adding New Data

The researcher will sometimes want to add variables from a new data source to existing cases in a file, or to add new cases to the file. Typical examples of the former are presented by follow-up or panel surveys where new information is collected on the same set of respondents on two or more occasions. Cases are matched on their identification numbers and the two files merged; if no information has been collected for a case on the second occasion it may either be excluded from the merged file or included but given missing values for the new variables.

Adding new cases to a data file is straightforward when the sequence of cases is of no significance: the new cases are merely appended. However if the data has a hierarchical structure new cases must be merged into the file at the appropriate points, using the identification field to make the match. Thus data on an additional member of a household would be grouped with data on other members of the household.

A special type of merging occurs when variables from a data set at a higher level of aggregation are added to a file. To illustrate, information on the constituency in which he lives may be needed for each member of a sample of voters, and the match is obtained on the constituency code. This operation is called a *duplicate merge* because the variables for a particular constituency must be duplicated for every sampled voter in the constituency.

OSIRIS is the only major package which has comprehensive facilities for merging data sets. Otherwise any new cases or variables must be appended to the data file and the whole file then re-sorted into order.

(d) Code Conversion

Any non-numeric codes in the data file should be converted to numeric format. Most programs will not accept non-numeric data, and those that do are more efficient if numeric data is inputted. The conversion of alpha-numeric codes is fairly straightforward: it can be done by a package which accepts such codes or, if this is not available, by a special purpose program in a high level language.

Multipunched codes are altogether a different proposition since very few packages can convert them and it is usually necessary for a computer specialist to write a routine in machine code. This routine should return for each multipunched card a matrix of 12 rows by 80 columns with the entry 1, if a hole is punched in a given cell, 0 otherwise. This gives complete information about the pattern of multipunches from which the numeric codes can be generated. The machine code routine can usually be incorporated in a high level language program but conversion is still a complicated and laborious process.

(e) Rectangularizing

Section 2 explained how rectangular matrices from surveys with a complex structure are derived. This operation is performed when data checking is complete and before analysis can proceed.

Different programs have very different requirements of data input. On the one hand the major packages designed for social surveys can accept data in a variety of formats and have comprehensive facilities for the generation of temporary variables at run time (although new variables which are used frequently are best added permanently to the file). In contrast many statistical programs require the data to be in the standard format of a complete data matrix with a fixed number of cards per case and numeric codes for all variables; some require data in a special form such as a contingency table. The packages thus have an important role to play in the 'bread and butter' work of survey analysis: operations such as the transformation and construction of variables, code conversion, the addition of new variables or cases, case selection and handling missing data.

A very useful feature in this respect is the *self-described file*, whereby the user supplies an additional file, known as the data dictionary. This contains such details as variable names and reference numbers, minimum and maximum values, variable type and location in the card image and file structure. The user can also include comments on the data file for his own use. The main advantage of the data dictionary is that it is linked to the data file at run time so that the user refers to variables by name or reference number and the data defining information is picked up automatically from the dictionary. The dictionary stores a full and up-to-date description of the file and it can be used to supply informative labels and headings for the results of analysis. A further feature of the package is that the data file and its associated dictionary can be stored in the computer in an internally compact form called *System* files rather than in space and time consuming card-image form. This leads to greater efficiency all round.

The major packages can calculate descriptive statisitics and perform basic statistical analyses such as multiple regression and factor analysis; OSIRIS also has some techniques particularly suited to survey data such as AID. As yet they do not incorporate many of the more recently developed techniques described in this book such as multidimensional scaling, the log-linear model and many types of cluster analysis. As we have seen, the programs available for these techniques do not have good data management facilities and are restrictive in their requirements of data input. Thus an important role for the packages is to provide the *interface* between the data file and these programs by creating work files in the required form. Both SPSS and OSIRIS allow the user to output selected cases and/or variables and correlation matrices in card image form from their system files for this purpose; in fact OSIRIS System files can be produced by SPSS and vice versa.

However the user may still find that there are problems of the interface with particular programs. Programs currently available for fitting log-linear models require a multiway table of frequencies as input. None of the major packages produce such tables in card-image form so that a laborious process of transfer from listing to cards must be carried out. Multidimensional scaling programs form another example, for the packages cannot produce most types of similarity/ dissimilarity matrices which the programs take as input.

A further aspect of the interface problem is that the data may have unusual features or the user may have requirements such as special data transformations which the program does not cater for. Normally a two phase operation is required in such cases – creating a work file in the required form by means of a special purpose program which is accessed by the analysis program in the second phase. Some computer operating systems allow these two phases to be executed without any intervention by the user. GSP is the only major package which avoids the two phase operation by allowing the user to incorporate his own *read-a-case* routine using the full power of a high level language to supply the variables for the analysis. This facility can get round many difficult problems and it is a pity that more packages and analysis programs do not have it. Similarly, incorporation of a user supplied routine for special output requirements would be an extremely useful facility.

2.6 STAGE 4: DESCRIPTION AND ANALYSIS

The characteristics of social survey data described in Section 2 lead to several special considerations at the final stage.

Calculating Sampling Errors

It is unfortunate that many packages and programs currently in use do not give sampling errors. Most of those that do have assumed simple random sampling, despite the fact that many surveys have a complex sample design. Recently this situation has been improved by the development of SEPP (Sampling Error Program Package) which calculates sampling errors for a comprehensive range of statistics including means, regression coefficients, ratios and correlation coefficients, for any complex sampling scheme, both for the whole sample and for subsamples. Further details are given in Chapter 1.

Use of Weights

Many packages allow the use of a weight variable which determines the extent to which each case contributes to the statistics being calculated. In cross-tabulation, for example, a case with weight 1.3 will result in 1.3 being added to both the total number of cases in the table and to the particular cell into which the case falls. Some statistical procedures such as rank-order correlation will not operate on fractional weights and most packages have *ad hoc* procedures for dealing with this. In SPSS for example, a weight of 3.6 would result in the reproduction of the case 3 times and a further time with probability 0.6. If weighting of cases is required but cannot be done in the program being used a simple but laborious solution is to reproduce each case w times in the data file (where w is the case weight and must be integer).

Use of Random Subsamples

Some packages have a facility for selecting a random sample of a specified proportion of cases from a file. This facility is extremely useful if the file has many cases, for analysis of a sample of cases may produce statistically significant results at far less cost. Another important use is to examine the stability of statistical analysis by comparing the results in a number of subsamples (see, for example, Chapters 1 and 8). The samples are usually chosen according to a repeatable sequence of random numbers. Ideally the user should be able to start at a chosen point in the sequence so that different subsamples of the same size can be obtained. Not all packages with the selection facility allow this.

Handling Missing Data

If missing values have not been imputed by one of the methods described in Section 4, cases with missing values must be deleted from the analysis. Different strategies of deletion are appropriate for different purposes.

(a) Univariate and bivariate statistics. In the calculation of simple statistics such as means or ratios, cases with missing values have to be excluded. However, some indication of the characteristics of the omitted cases helps the researcher to assess the bias thus introduced into the estimates. Siegle (1967) describes some procedures for the evaluation of the non-random effects of missing data on the estimates of means, variances and the t-test for means. With categorical variables a missing data code can be of interest in itself, particularly if it represents a refusal or 'don't know'. Therefore all major packages have the option of either including or excluding cases with missing values in cross-tabulation.

(b) Techniques based on a correlation matrix. These include multiple regression, factor analysis, principal components, partial correlation, multidimensional scaling and cluster analysis. For any particular analysis we have p variables X_{1k}, X_{2k},

. . . ,X_{pk} for the kth case, one or more of which may be missing. The correlation between variables i and j is

$$r_{ij} = \frac{\sum_{k=1}^{n} (x_{ik} - \bar{x}_i)(x_{jk} - \bar{x}_j)}{\left[\sum_{k=1}^{n} (x_{ik} - \bar{x}_i)^2 \sum_{k=1}^{n} (x_{jk} - \bar{x}_j)^2\right]^{1/2}}$$

It is obvious that if either X_{ik} or X_{jk} is missing no contribution can be made to the correlation coefficient r_{ij} by case k. There are two ways of dealing with this situation. The first, called *listwise deletion*, is to omit entirely any case with at least one missing value from the calculation of the correlation matrix. The greater the number of variables in the analysis and the more missing values are spread among cases, the greater will be the reduction in the number of cases on which the analysis is based. The second method, *pairwise deletion*, omits a case from the calculation of a particular coefficient if one or both values are missing, but includes it for the calculation of coefficients where both values are present. This method thus uses as much data as possible but each coefficent may be based on a different 'n'. Great care is required in the interpretation of analyses based on the resulting matrix because different coefficients may be based on different parts of the sample. In contrast the listwise deletion method gives the same subsample for analysis, though it may be a small and biased one.

Listwise deletion is fairly straightforward to carry out with most analysis programs since all that is required is a facility for selecting cases for analysis on the basis of say a variable that has been set to 1 if there are no missing values, 0 otherwise. However special facilities are required for pairwise deletion and these are incorporated in most major packages but not in individual analysis programs. If such facilities are not available, pairwise deletion can be effected by assembling the correlation matrix from coefficients that have been calculated separately.

(c) Techniques based on a distance matrix. These include multidimensional scaling and cluster analysis. Again we have variables $X_{1k}, X_{2k}, \ldots, X_{pk}$ for the kth case but the coefficient is based on an averaging over variables rather than cases. An example of a dissimilarity coefficient for cases k and l based on euclidean distance is

$$d_{kl} = \frac{1}{p}\left\{\sum_{i=1}^{p} (x_{ik} - x_{il})^2\right\}^{1/2}$$

The treatment of missing values is analogous to that for statistics based on a correlation matrix except that we have variable deletion instead of case deletion. Listwise deletion gives distance coefficients calculated on the same (restricted) set of variables while pairwise deletion may result in the coefficients for different pairs of cases being based on different variables.

It is very difficult to give any general guidance on the treatment of missing values in survey analysis. Both imputation and case deletion are in some respects *ad hoc* and the choice of method must depend upon the researcher's knowledge of the data and the assumptions he is prepared to make.

2.7 APPENDIX

Some Major Packages for Survey Processing and Analysis

SPSS [Statistical Package for the Social Sciences], Nie et al (1975)

A popular and well-documented integrated system of programs with comprehensive facilities for data handling and description. Its main disadvantages are that it will only handle files with a fixed number of cards per case and does not have good file building facilities, although a new version soon to be released will handle complex structures and has extended analysis facilities. It is available on a wide range of computers and in most universities.

OSIRIS [Organized System of Integrated Research Investigation Statistics], OSIRIS III (1973) and Rattenbury & Pelletier (1974)

A large and comprehensive collection of programs covering the whole range of facilities required for survey processing and analysis, including file creation and maintenance for complex as well as simple file structures. It has a number of analysis programs specially designed for surveys. Its main drawbacks are that it is available, with few exceptions, only on IBM computers in the USA and is relatively complicated to use.

BMD [Bio-Medical Computer Programs], Dixon (1971)

A large collection of statistical programs which is available on many computers. It has limited data handling and description capabilities and is complicated to use but has the widest range of statistical analyses of all the packages.

SEPP [Sampling Error Program Package], Kish *et al.* (1972)

The only generally available program which calculates sampling errors for complex designs.

GSP [General Survey Program], Yates (1973)

This is not yet widely available but has some very useful facilities not contained in the other packages. In particular it has flexible input requirements (the user supplies his own Fortran routine) and powerful facilities for the manipulation and presentation of tables of such quantities as frequencies, means and ratios.

P-STAT (R Buhler, Computer Centre, University of Princeton, New Jersey)

A comprehensive package for data processing and analysis which is especially good for file management and the handling of complex data structures. Available on a number of computers but mainly confined to North America.

REFERENCES

Anderson, R. E. and Coover, E. R. (1972), 'Wrapping up the package: critical thoughts on applications software for social data analysis', *Computers and the Humanities*, **7**, 81–95.

Dixon, W. J. (ed.) (1971), *BMD-Biomedical Computer Programs*, Berkeley: Also, (1970), P Series Supplement, University of California Press.

Elashoff, R. M. and Afifi, A. (1966), 'Missing values in multivariate statistics – I, review of the literature', *J. Am. Statist. Ass.*, **315**, 595–604.

Freund, R. J. and Hartley, H. O. (1967), 'A procedure for automatic data editing', *J. Am. Statist. Ass.*, **62**, 341–352.

Goldstein, H. (1970), 'Data processing for longitudinal studies', *Appl. Statist.*, **19**, 145–151.

Hocking, R. R., Huddleston, H. F. and Hunt, H. H. (1974), 'A procedure for editing survey data', *Appl. Statist.*, **23**, 121–133.

Janda, K. (1969), *Data Processing Applications to Political Research*, 2nd Edition, Evanston: Northwestern University Press.

Kish, L., Frankel, M. R., and Van Eck, N. (1972), *SEPP-Sampling Error Program Package*, Ann Arbor: Institute for Social Research, University of Michigan.

Moser, C. A. and Kalton, G. (1971), *Survey Methods in Social Investigation*, 2nd Edn., London: Heinemann.

Nie, N. H., Bent, Dale H, and Hull, Hadlai C. (1975), *Statistical Package for the Social Sciences*, 2nd Edn., New York: McGraw-Hill.

OSIRIS III (1973) Volume 1, Ann Arbor: Institute for Social Research, University of Michigan.

Rattenbury, J. and Pelletier, R. (1974), *Data Processing in the Social Sciences with OSIRIS*, Ann Arbor: Institute for Social Research.

Richardson, M. and Cunningham, A. D. (1972), 'Vetting of industrial survey questionnaires by computer', *Appl. Statist.*, **21**, 53–54.

Siegle, D. (1967), 'Some aids in handling missing data', *Social Science Information*, **5**, 133–150.

SSRC (1973, 1974, 1975), *Computer Software for Survey Analysis*, Special Supplements in SSRC Newsletters 20, 25, and 29, London: Social Science Research Council.

Yates, F. (1973), 'The Analysis of Surveys on Computers – Features of the Rothampstead Survey Program', *Appl. Statist.*, **22**, 161–171.

Chapter 3

Cluster Analysis

B. S. Everitt

3.1 PRINCIPLES OF CLUSTER ANALYSIS

3.1.1 INTRODUCTION

The techniques of *cluster analysis*, of which there are many, are used in the exploration of data that arise from the measurement of a number of characteristics for each of an assorted collection of individuals or objects. The aim of the exploration is to determine if the latter may be subdivided into groups or clusters which, on the basis of the measurements, can be shown to be relatively distinct or to belong together. The aim is clearly different from that of *discriminant function analysis*, or similar *assignment* techniques, used for the allocation of people to known groups. Unlike these, cluster analysis is concerned with the more difficult and intrinsically more interesting problem of discovering the groups in the first instance.

A fundamental problem in this area is the lack of a satisfactory definition of exactly what constitutes a cluster. Because of this, most clustering techniques cannot be formulated in terms of a satisfactory model as can, for example, principal components or factor analysis (see Chapter 4). The problem of definition has been considered by several authors, for example, Ling (1972), and is obviously crucial in many respects. However, since the main aim of this chapter is merely to acquaint readers with some particular clustering techniques and their associated problems, the matter of definition will conveniently be ignored, and the term *cluster* or *group* used in an intuitive sense for a collection of 'similar' individuals or objects.

The reader should bear in mind throughout this chapter that most cluster analysis methods are essentially *non-statistical* in the sense that they have no associated distribution theory or significance tests, and so are unable to relate from sample to population. Indeed many clustering techniques treat the data at hand *as* the population and so there is no question of making inferences as to the underlying structure in some hypothetical population from which the data was drawn. They are perhaps best regarded as techniques for generating hypotheses rather than methods for testing hypotheses.

3.1.2 Some Possible Reasons For Using Clustering

Investigators may be interested in applying clustering techniques to their data for several reasons. The first reason could be as a means of data description or summarization. It is easy to give examples (see Ball, 1971), where very dissimilar data sets have the same mean and variance–covariance matrix, and where quoting these statistics would not be a useful summary of the data. Dividing the data into clusters and quoting the *within* cluster statistics may well be a far more meaningful description in many cases. A second reason for clustering may be for data reduction, using the techniques to reduce the volume of a large set of data by substituting for the n original data cases (where n is large perhaps of the order of 5,000 – 10,000) a number, say g, of cluster means (where g is relatively small, say of the order of 150 – 200) which will tend to characterize the variation of the data and provide a relatively accurate summary. This may be useful for presentation or communication purposes, and also useful as the starting point for further multivariate analyses, such as factor analysis or multidimensional scaling (see Chapter 6) which are not able to handle very large data sets. A further reason for using cluster analysis, and perhaps the one which most investigators see as the prime reason, is that of attempting to discover whether their data reflects some true typology in their field of study, in the sense of containing well separated clusters which arise from some basic fact of nature as, for example, does the man-women classification.

In the case of data description and data reduction, clustering is being used primarily as a method of obtaining a convenient summary of one's data. If some true typology is suspected, discovery of the groups is a necessity since they will be central to both future research and theoretical development in the area. This distinction appears to be similar to that made by Kendall (1966), who calls the situations *dissection* and *classification* respectively. In the first case individuals in one group are not required to be distant from individuals in another group, whereas in the second case individuals belonging to different groups *are* required to be distant from each other. Cormack (1971) makes the point that all collections of individuals can be dissected but not all can be classified.

Since many users of these techniques are unclear *a priori* whether dissection or classification is their reason for clustering, a problem arises once a set of clusters have been obtained; should they be regarded as indicating some true typology and labelled accordingly? (Subjectivity in the naming of clusters is probably of the same magnitude as subjectivity in the naming of factors); alternatively should they be regarded merely as *administrative* clusters which are useful as a meaningful summary of the data. Such questions are probably best answered by careful evaluation of clusters in terms of predictive ability, etc.: a matter which is discussed more fully in Section 3.2.5.

3.1.3 Similarity and Distance Measures

The input to the majority of clustering techniques consists of an $n \times p$ matrix of measurements, say X, where

$$\mathbf{X} = \begin{bmatrix} x_{11} & x_{12} & \cdot & \cdot & x_{1p} \\ x_{21} & & & & \\ x_{n1} & & \cdot & \cdot & x_{np} \end{bmatrix},$$

in which x_{ij} is the value taken by individual i on variable j. The first stage in many methods of cluster analysis is the conversion of this matrix into an $n \times n$ matrix of inter-individual *similarities*, *dissimilarities* or *distances*. These are terms for measures of the relationship between individuals, given the values of a set of p variates common to both. Full accounts of such measures are given in Sokal and Sneath, (1963), Williams and Dale, (1965), and Everitt, (1974). In this section a brief description only is given for those readers totally unfamiliar with the area. In most of what follows we shall assume that we are interested in clustering the 'individuals' although many of the techniques to be discussed would be equally suitable for clustering 'variables'. For clustering variables we would convert the matrix of observations into a $p \times p$ matrix of inter-variable similarities.

(1) Similarity Measures. Similarity coefficients take values in the range zero to one, with high values indicating that the two individuals are alike with respect to the p variables measured. Perhaps the most commonly used similarity coefficient is the product moment correlation coefficient, although this is probably more well known as a measure of the relationship between variables. Many other similarity coefficients are known, for example, the simple matching coefficient for use with binary data, which defines the similarity between two individuals, as the number of matches divided by the total number of variables. Lists or other similarity coefficients are given in the references previously mentioned.

(2) Distance Measures. Distance measures may in general take any positive value with high values indicating that the individuals tend to be unalike with respect to the p variables measured. That is high inter-individual distances indicate low inter-individual similarity. Distance measures, are more formally defined below. Perhaps the most commonly used distance measure is Euclidean distance which is also given below.

Distance

(1) Distance Measures. A numerical function $d(x,y)$ of pairs of points of a set E is said to be a *distance function* or *a metric* for E if it satisfies the following conditions:-

(a) $d(x,y) \geqslant 0$; $d(x,y) = 0$ if $x = y$;
(b) $d(x,y) = d(y,x)$;
(c) $d(x,z) + d(y,z) \geqslant d(x,y)$;

The third condition is the one which differentiates most between distance measures and similarity measures. This condition is referred to as the *metric inequality*, or more commonly as the *triangular inequality*.

(2) Euclidean Distance. Probably the most commonly used distance measure, and the most familiar, is the Euclidean metric, where the distance between points i and

j, d_{ij}, is defined as

$$d_{ij} = \left\{ \sum_{k=1}^{p} (x_{ik} - x_{jk})^2 \right\}^{1/2},$$

where x_{ik} is the value on the kth variable for the ith individual.

3.1.4 An Example of the Operation of a Clustering Technique

For those readers completely unfamiliar with clustering techniques, a brief numerical example of the operation of a particular method will be given in this section in the hope that this will enable such readers to follow more clearly the rest of the chapter.

Let us suppose we wish to cluster the following set of data in which four observations are recorded for each of six individuals.

$$\mathbf{X} = \begin{bmatrix} 6.0 & 3.0 & 4.0 & 5.0 \\ 2.0 & 3.0 & 5.0 & 4.0 \\ 5.0 & 4.0 & 6.0 & 3.0 \\ 9.0 & 1.0 & 1.0 & 8.0 \\ 8.0 & 2.0 & 0.0 & 9.0 \\ 8.0 & 0.0 & 1.0 & 8.0 \end{bmatrix}$$

The clustering method to be used is known as *group average* clustering, and the first stage is to compute an interindividual similarity or distance matrix. We shall in this example use Euclidean distances between individuals, and these are shown in the matrix $\mathbf{D}_1$.

$$\mathbf{D}_1 = \begin{bmatrix} 0.0 & 4.24 & 3.16 & 5.57 & 6.08 & 5.57 \\ & 0.0 & 3.46 & 9.22 & 9.33 & 8.77 \\ & & 0.0 & 8.66 & 9.22 & 8.66 \\ & & & 0.0 & 2.00 & 1.41 \\ & & & & 0.0 & 1.41 \\ & & & & & 0.0 \end{bmatrix},$$

(Since the matrix is symmetric only the entries above the diagonal are given). For example, the distance between individuals one and two is computed as follows:-

$$\begin{aligned} d_{12} &= [(6.0 - 2.0)^2 + (3.0 - 3.0)^2 + (4.0 - 5.0)^2 + (5.0 - 4.0)^2]^{1/2} \\ &= 4.24. \end{aligned}$$

Group average clustering is a particular example of a *hierarchial* method (see Section 3.2). It begins by assuming that each individual is a single member cluster, and then at each stage reduces the number of clusters by one by joining together individuals or clusters of individuals, until, at the final stage, all individuals are in a single group. The problem of deciding which solution, that is which set of clusters, to accept is discussed in Section 2.3. We shall now see how this method operates on the set of data given above.

The first step in the clustering is to form a group of the two individuals closest together. Examination of $\mathbf{D}_1$ shows that the smallest entries are d_{46} and d_{56} both of which equal 1.41. Either of these pairs might be taken to form a group, so we shall take individuals 5 and 6. Therefore, at this stage we have five clusters, four of which contain a single individual and one of which contains individuals 5 and 6.

The next stage is to compute a new distance matrix showing interindividual distances and individual to cluster distances. Group average clustering defines the distance between an individual and a cluster containing several individuals as the average of the distances between the individual and the members of the cluster. For example, the distance between individual 1 and the cluster containing individuals 5 and 6 is defined as

$$
\begin{aligned}
d_{1,(56)} &= \tfrac{1}{2}(d_{15} + d_{16}) \\
&= \tfrac{1}{2}(6.08 + 5.57) = 5.83.
\end{aligned}
$$

We therefore arrive at the following distance matrix:

$$
\mathbf{D}_2 = \begin{array}{c} \\ 1 \\ 2 \\ 3 \\ 4 \\ (56) \end{array}
\begin{array}{c}
\begin{array}{ccccc} 1 & 2 & 3 & 4 & (56) \end{array} \\
\left[\begin{array}{ccccc}
0.0 & 4.24 & 3.16 & 5.56 & 5.83 \\
 & 0.0 & 3.46 & 9.22 & 9.05 \\
 & & 0.0 & 8.66 & 8.94 \\
 & & & 0.0 & 1.71 \\
 & & & & 0.0
\end{array}\right]
\end{array}
$$

The smallest entry in this matrix is 1.71, which is the distance between individual 4 and cluster containing individuals 5 and 6. Therefore at this stage individual 4 joins this cluster, and a further distance matrix, $\mathbf{D}_3$, is computed.

$$
\mathbf{D}_3 = \begin{array}{c} \\ 1 \\ 2 \\ 3 \\ (456) \end{array}
\begin{array}{c}
\begin{array}{cccc} 1 & 2 & 3 & (456) \end{array} \\
\left[\begin{array}{cccc}
0.0 & 4.24 & 3.16 & 5.74 \\
 & 0.0 & 3.46 & 9.11 \\
 & & 0.0 & 8.85 \\
 & & & 0.0
\end{array}\right]
\end{array}
$$

The smallest entry in $\mathbf{D}_3$ is 3.16, the distance between individuals 1 and 3 and so these are now joined. At this stage therefore we have three clusters namely

	Cluster 1	Cluster 2	Cluster 3
Members	2	1,3	4,5,6

A new distance matrix $\mathbf{D}_4$ is now computed. In this matrix we need the distance between two clusters (clusters 2 and 3), both of which contain several individuals. Group average clustering defines this distance as the average of the distances between all pairs of individuals in the two groups. Therefore the distance between clusters 2 and 3 is given by

$$
\begin{aligned}
d_{(23)(456)} &= {}^{1}/_{6}(d_{24} + d_{25} + d_{26} + d_{34} + d_{35} + d_{36}) \\
&= {}^{1}/_{6}(9.22 + 9.33 + 8.77 + 8.66 + 9.22 + 8.66) \\
&= 8.98
\end{aligned}
$$

$\mathbf{D}_4$ is therefore as follows:

$$
\mathbf{D}_4 = \begin{array}{c} 2 \\ (13) \\ (456) \end{array}
\overset{\begin{array}{ccc} 2 & (13) & (456) \end{array}}{\begin{bmatrix} 0.0 & 3.85 & 5.74 \\ & 0.0 & 8.98 \\ & & 0.0 \end{bmatrix}}
$$

The smallest entry in $\mathbf{D}_4$ is 3.85, which is the distance between individual 2 and the cluster containing individuals 1 and 3. Therefore individual 2 joins this cluster, and at this stage we have the following two clusters:

	Cluster 1	Cluster 2
Members	1,2,3	4,5,6

The last stage is the fusion of these two clusters. The series of results presented here may be shown diagrammatically as a *dendrogram*, which is a diagram showing the fusions of individuals taking place at each stage of the procedure. The dendrogram for this example is shown in Figure 3.1.

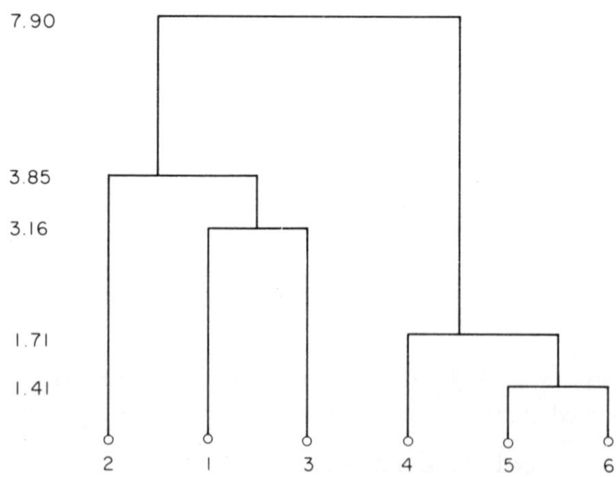

Figure 3.1. Group average clustering dendrogram.

3.2 CLUSTERING PROCEDURES

3.2.1 Choice of Variables

The basic data for a cluster analysis is a set of n individuals or objects on which have been recorded a number, p, of measurements. This initial choice of the particular set of measurements used to describe each individual, constitutes a frame of reference within which to establish the clusters and the choice reflects, to some extent, the investigators judgement of relevance for the purpose of clustering. Some attention needs to be given, therefore, to asking whether the variables chosen are relevant to the type of classification being sought. It is easy to think of situations in which the inclusion of 'irrelevant' variables may produce clusters which are, at the least, uninformative, for the purpose at hand. For example, if a classification of the mentally ill is sought for the purpose of investigating the effect of different treatments on different types of patient, it is, in general, not sensible to include such variables as height, weight, and other 'vital' statistics, since on the basis of these variables the most likely classification is into males and females, a 'useful' classification in other respects, but almost certainly not for this purpose.

In many areas of research the data collected will consist of a 'mixture' of variables; some will be of the 'presence' and 'absence' variety, some will be categorical with several possible categories, and yet others will be interval level or 'quasi' interval level. Applying cluster analysis to such data is a problem since many techniques are not suitable for use with such a variety of different types of measurement. In some situations it may be suitable to consider the variables separately, and to cluster the data using first the qualitative variables, and then to repeat the analysis using only the quantitative variables. If, however, it was desired to analyse the data using *all* the variables one possibility would be to first compute the interindividual similarity matrix using Gower's similarity coefficient (see Gower, 1971), and then use one of the many clustering techniques which operate on similarity matrices, to produce a set of clusters. Gower's similarity coefficient is particularly suitable here since it can deal with a mixture of the various types of variables.

In many cases it may be useful to reduce, in some way, the number of variables for each individual before applying any particular clustering technique. A simple was to do this, and one which has been used in several studies (for example, Everitt *et al* (1971)), is to perform a principal components analysis on the data, and then use the first few principal component scores for each individual as input to the method of cluster analysis to be employed.

3.2.2 A Brief Review of Clustering Techniques

Comprehensive reviews of clustering techniques are given by Cormack (1971) and by Everitt (1974). In this section some of the better known methods are briefly described.

(1) Hierarchial Techniques

Hierarchial techniques are probably the most commonly used class of clustering methods. They may be subdivided into *agglomerative* methods which proceed by a series of successive fusions of the n individuals into groups, and *divisive* methods which partition a set of n individuals successively into finer partitions.

Agglomerative Methods. We have already seen in Section 3.1.4 how one particular hierarchial agglomerative technique, namely group average clustering, operates. The basic procedure with all other methods in this class is similar. They begin with the computation of an interindividual similarity or distance matrix, and finish with a dendrogram showing the results of the clustering. At any particular stage the methods fuse individuals or clusters of individuals which are closest (or most similar). Differences between methods arise because of the different ways of defining distance (or similarity) between an individual and a cluster containing several individuals, or between two clusters. In the case of the group average methods we have seen that distance between clusters is defined as the average distance between all pairs of individuals in the two clusters. Other methods define this distance in other ways. For example, single linkage clustering defines it as the distance between the two clusters closest members, whilst centroid clustering defines distance between clusters as the distance between clusters centroids. A general formula for the distance between clusters used by many of these agglomerative techniques is given below.

Divisive Methods. With divisive methods the first task is to split the initial set of individuals into two. Now a set of n individuals can be divided into two sub-sets in $2^{n-1}-1$ ways, and Edwards and Cavalli-Sforza (1965) consider all possible splits and pick that with minimum within group sum of squares. This is obviously only possible for very small values of n even with a powerful computer, since the numbers evolved are enormous. (Gower (1967) calculates that to consider all splits for $n = 41$ would take in the region of 500 *years* of computer time even using the fastest machines available). In the case of even moderate sized sets of data a restriction must be imposed on the number of ways considered.

MacNaughton-Smith *et al.* (1964) describe a divisive technique for which the computation involved is considerably less than for 'an all possible sub-divisions method'. With this method a splinter group is accumulated by sequential addition of the individual whose total dissimilarity with the remainder less its total dissimilarity with the splinter group is a maximum. When this difference becomes negative the process is repeated, if desired, on the two sub-groups. The measure of dissimilarity used is the average Euclidean distance between an individual and the other individuals in the group.

Perhaps the most well-known divisive techniques are the *monothetic* techniques, which are used in situations involving binary data. A division of the data is then initially into those individuals who possess and those who lack some one specified attribute, say variable T. That is, one subset will contain those objects which possess T and the other individuals which lack T. Variable T is chosen so that dissimilarity between the two clusters is maximized in terms of the selected criterion. This type of technique is usually known as *association analysis*, and several variants of the technique have been given, for example Lambert and Williams (1966) and MacNaughton-Smith (1965) differing in the division criterion adopted.

(2) Optimization Methods

Another major class of clustering techniques are those which seek to partition the set of data so as to optimize some predefined clustering criterion. They differ from the hierarchical techniques discussed previously in that they admit relocation of the individuals thus allowing poor initial partitions to be corrected at some later stage of the analysis. These techniques assume that the number of groups into which to partition the data has been set by the user to some value say k (see Section 3.2.2 for a discussion of methods for deciding on the best value of k).

Many clustering criteria have been suggested. Probably the most common is minimization of the pooled within cluster sum of squares, a method proposed by several authors, for example, Singleton and Kautz (1966) MacQueen (1966) and Ball and Hall (1967). The rationale behind this criterion is that of seeking compact homogeneous clusters, described by Cattell and Coulter (1966), as *homostats*. Several other clustering criteria have been suggested, some of which are described below.

The obvious way to optimize any cluster criterion is to consider *every* partition of the n objects into k clusters, and pick that partition with optimal criterion value. Unfortunately the number of ways of partitioning n objects into k non-empty subsets is extremely large (see below), and consequently total enumeration would only be feasible for very small n and is never used in practice. In general, some type of iterative 'k-means' approach is used in which, starting with an initial partition into k groups (which can be obtained in several ways, or could be supplied by the investigator), each individual is considered for moving from the group it occupies to each other group, the move being made if it causes an increase (or decrease in the case of minimization), in the value of the cluster criterion. This process is continued until no move of a single individual causes an improvement in the criterion value. This solution can be accepted or an attempt could be made to reach an improved optima using some different starting configuration.

(3) Other Clustering Techniques

Wolfe in a series of papers (1967, 1969, 1970) attempts to formulate a more statistical approach to cluster analysis. He assumes that the data are a sample from some population in which the distribution of the variables takes a particular form. The form assumed is a mixture of multivariate normal distributions, and arises from assuming that each cluster has a different multivariate normal distribution for the variables. The problem is to identify and describe the cluster distributions from a sample drawn from the mixture. Once these have been identified, the probabilities of individuals belonging to each of the clusters may be found, and the maximum of these may be used to associate each individual with a particular cluster. This approach may be formulated in concise mathematical terms (see technical section below), and the identification of clusters becomes a problem of estimation. For this, Wolfe uses maximum likelihood techniques.

If the data consisted of measurements on two or three variables, it could be examined visually to determine whether clusters were present. Unfortunately this is not generally the case. There are however ordination techniques which allow the mapping of the original p-dimensional data into a lower dimensional space, say p^* ($p^* < p$). Hopefully, in many cases, a useful representation may be obtained with $p^* = 2$ enabling a plot of the data to be made. Perhaps the most common ordination technique is principal components analysis. A plot of the data in the

Figure 3.2. Faces representing thirty five dimensional observations.

space of the first two or perhaps three components may give some indication of the structure present. Similarly the scaling techniques of Kruskal (1964) and others, could be used to obtain a two dimensional mapping of the data. Sammon (1969) and Chang and Lee (1973) describe a technique called *Non-Linear Mapping*, which maps the p-dimensional data into p^* by attempting to minimize an expression D (see below) which is a function of the distances between points in the original p-dimensional space and the distances between points in the reduced p^* space. Other methods for obtaining graphical representation of multivariate data are given by Andrews (1972) and Ling (1972).

A recent paper by Chernoff (1973) describes a method which may be useful for investigating data for possible clusters. In this method each p-dimensional observation is represented by a cartoon face, the characteristics of the face being determined by the values of the p variables. A sample of observations is represented by a collection of faces, and a 'cluster analysis' of the data may be performed by grouping similar faces. For example, Figure 3.2, shows a collection of thirty faces which represent thirty five-dimensional observations. These observations were generated to contain three well-separated clusters, and this structure is shown clearly by the faces.

The representation of the data in terms of faces may be more useful than others since people are used to studying and reacting to faces, and hopefully they are able to filter out insignificant visual phenomena and focus on the potentially important. Certain major characteristics of the faces are instantly observed and finer details became apparant after studying the faces for some time. The awareness of these does not drive out of mind the original major impressions.

A comparison of Sammon's Non-Linear Mapping technique, Andrews' multi-dimensional plotting method and Chernoff's 'faces', for representing multivariate data, is made in Everitt and Nicholls (1975).

(4) *Technical Section*

(a) Between cluster distance measures for hierarchial agglomerative techniques. The distance measures between clusters used by many hierarchial agglomerative techniques satisfy the following recurrence formula.

$$d_{k(ij)} = \alpha_i d_{ki} + \alpha_j d_{kj} + \beta d_{ij} + \gamma \mid d_{ki} - d_{kj} \mid ,$$

where $d_{k(ij)}$ is the distance between a cluster k and a cluster ij formed by the fusion of clusters i and j, and d_{ij} is the distance between clusters i and j. α, β, and γ are parameters whose values for various methods are as follows:

(i) Single Linkage: $\alpha_i = \alpha_j = \frac{1}{2}; \beta = 0; \gamma = -\frac{1}{2}$:

(ii) Centroid Clustering: $\alpha_i = n_i/(n_i + n_j); \alpha_j = n_j/(n_i + n_j); \beta = -\alpha_i\alpha_j; \gamma = 0$

(iii) Group Average: $\alpha_i = n_i/(n_i + n_j); \alpha_j = n_j/(n_i + n_j); \beta = \gamma = 0$

n_i is the number of individuals in cluster i and n_j is the number of individuals in cluster j.

(b) Some Division Criteria for Association Analysis. Suppose our n individuals, have been scored on p 'present-absent' attributes, and a, b, c, and d are all the usual cell counts in a 2 x 2 table for attributes j and k. A $p \times p$ matrix of chi-squares may be obtained, where

$$\chi^2_{jk} = \frac{(ad - bc)^2 n}{(a + b)(a + c)(b + d)(c + d)}$$

If we suppose that division of the data is on attribute k, into those individuals possessing this attribute and those who do not, criteria which have been suggested for choosing k are as follows:

$$\text{Max}\left[\sum_{j \neq k}^{p} x^2_{jk}\right]$$

$$\text{Max}\left[\sum_{j \neq k}^{p} \sqrt{(x^2_{jk})}\right]$$

$$\text{Max}[\Sigma \mid ad - bc \mid]$$

$$\text{Max}[\Sigma(ad - bc)^2]$$

(c) Clustering Criterion

(a) Minimization of det (**W**)

(b) Maximization of trace ($\mathbf{BW}^{-1}$).

where W is the within groups matrix of sums of squares and cross products and B is the corresponding between groups matrix. Both of these criteria were suggested by Friedman and Rubin (1967).

(d) Number of Partitions of n objects into k non-empty sub-sets. Fortier and Solomon (1966) have considered this problem and they give the following recurrence formula.

$$P(n, k) = (k^n - \sum_{i=1}^{k-1} k_{(k-i)} P(n, i))/k!),$$

where $k \geqslant 2$; $n \geqslant k$. $P(n,k)$ is the number of distinct partitions of n objects containing exactly k non-empty subsets, and $k_{(k-i)} = k(k-1) \ldots (k - i + 1)$. For example, with $n = 19$ and $k = 8$ there are 1,709,751,003,480 distinct partitions.

(e) Fitting a Mixture of Multivariate Normal Distributions (Wolfe's Clustering method). Wolfe assumes that the data comes from a population having the distribution $f(\mathbf{x})$ given by

$$f(\mathbf{x}) = \sum_{s=1}^{g} \lambda_s \alpha_s(\mathbf{x}, \boldsymbol{\mu}_s, \Sigma_s)$$

where $0 \leqslant \lambda_s \leqslant 1$ $\sum_{s=1}^{g} \lambda_s = 1$, g = number of clusters, and

$$\alpha_s(\mathbf{x}, \boldsymbol{\mu}_s, \boldsymbol{\Sigma}_s) = (2\pi)^{-p/2} \, |\boldsymbol{\Sigma}_s|^{-\frac{1}{2}} \exp -\tfrac{1}{2}(\mathbf{x} - \boldsymbol{\mu}_s)' \boldsymbol{\Sigma}_s^{-1} (\mathbf{x} - \boldsymbol{\mu}_s)$$

Estimates of the mixing proportions λ_s the mean vectors μ_s, and the covariance matrices Σ_s are obtained by maximum likelihood methods. Each component multivariate normal distribution represents a separate cluster and individuals are then assigned to the cluster to which their probability of belonging is greatest. This probability is given by

$$P(s/\mathbf{x}) = \frac{\lambda_s \alpha_s(\mathbf{x}, \mu_s \Sigma_s)}{f(\mathbf{x})},$$

where $P(s/\mathbf{x})$ is the probability of an individual with vector of scores $\mathbf{x}$ belonging to cluster s. s may take values from 1 to g.

(f) Non-Linear Mapping. This technique obtains a lower dimensional mapping by seeking to minimize the expression D given by

$$D = \frac{1}{\sum\limits_{i<j} d_{ij}} \sum_{i<j} \frac{(d_{ij} - d_{ij}^*)^2}{d_{ij}},$$

where the d_{ij}'s are the distances between points i and j in the original p-dimensional space and the d_{ij}^*'s are the distances in the lower p^*-dimensional space. The procedure starts by obtaining an arbitrary two-dimensional configuration (assuming we are interested in the $p^* = 2$ solution), for example, by using the two original dimensions with largest variance, and then proceeds to adjust points so that D is minimized. Minimizing the expression D tends to preserve local structure in the sense that each point in p^*-space tends to bear approximately the same relationship to its neighbours as it did in p-space, while its relationship to points relatively distant in p-space may be quite different in p^*-space.

3.2.3 A Brief Description of some Problems of Clustering Techniques

There are many problems associated with using clustering techniques in practice; of these perhaps the most important is in the area of evaluating solutions, which will be discussed in Section 3.2.5. In this section some of the more technical problems of these techniques are briefly described; a fuller description is available in Everitt (1974, Chapter 3).

(1) Number of Clusters

A major problem when using clustering techniques is that of deciding on a particular number of clusters which best fits the data. Various suggestions have been made in this area. One is to examine dendrograms for large differences between successive nodes, and to take this as indicating which number of clusters to consider. When using one of the optimization methods described in Section 3.2.2 a procedure sometimes advocated is to plot the value of the clustering criterion against number of groups and to examine the plot for sharp changes in going from $g-1$, to g groups. This would then indicate that the g cluster solution should be considered the most satisfactory. Most of these procedures can be shown to have very little value. (For examples, see Everitt, 1974). The problem is of course, as mentioned in the introduction that there is no satisfactory definition of exactly what constitutes a cluster, and formal assessment of the significance of clusters could only be an analytic problem relative to a particular definition. In practice solutions for a range of values of g are examined, and the clusters judged in terms of interpretability and predictive ability (see later).

(2) Suboptimal Solutions

For those techniques which form clusters by seeking to optimize some criterion (see Section 3.2.2), there is the problem, already mentioned, that total enumeration of all possible partitions is not feasible, and consequently an algorithm must be used which may not find the solution with *global optimum*, but instead may find one with a *local optimum* of the clustering criterion. Consequently when using these techniques in practice several runs should be made, using in each case a different initial configuration. This may be obtained from some simpler clustering technique or may be supplied by the user in some cases. The values taken by the clustering criterion for the various solutions may then be compared, and that solution accepted which has the highest (lowest in the case of minimization) criterion value.

(3) Imposing a Structure on the Data

Many clustering techniques make implicit assumptions about the structure present in the data, and in cases where such assumptions are not met they may *impose* a structure rather than discover the actual structure present. An example of this is those techniques which are biased towards producing spherical clusters. (For example, methods such as centroid clustering and minimization of the within clusters sum of squares). If such methods are applied to data containing clusters of other shapes misleading solutions may result. For example, Figure 3.3 shows a set of data containing two well separated elliptical clusters. A method of cluster analysis designed to find groups by minimizing the within cluster sum of squares applied to this data gives the solution shown in Figure 3.4 (The 'ones' in the plot represent the observations grouped together by this method into the first cluster, the 'twos' those observations in the second). Figure 3.4 illustrates clearly what has happened. This method has *imposed* a spherical cluster structure on the data and has, therefore, not found the elliptical groups actually present. Further examples are given in Everitt (1974). A possible way to help avoid spurious solutions due to this, or other, causes is to apply several different clustering techniques to one's data, and to compare the clusters produced by different methods. Clusters found only by particular techniques might be regarded as suspect, until other evidence for their existence can be produced.

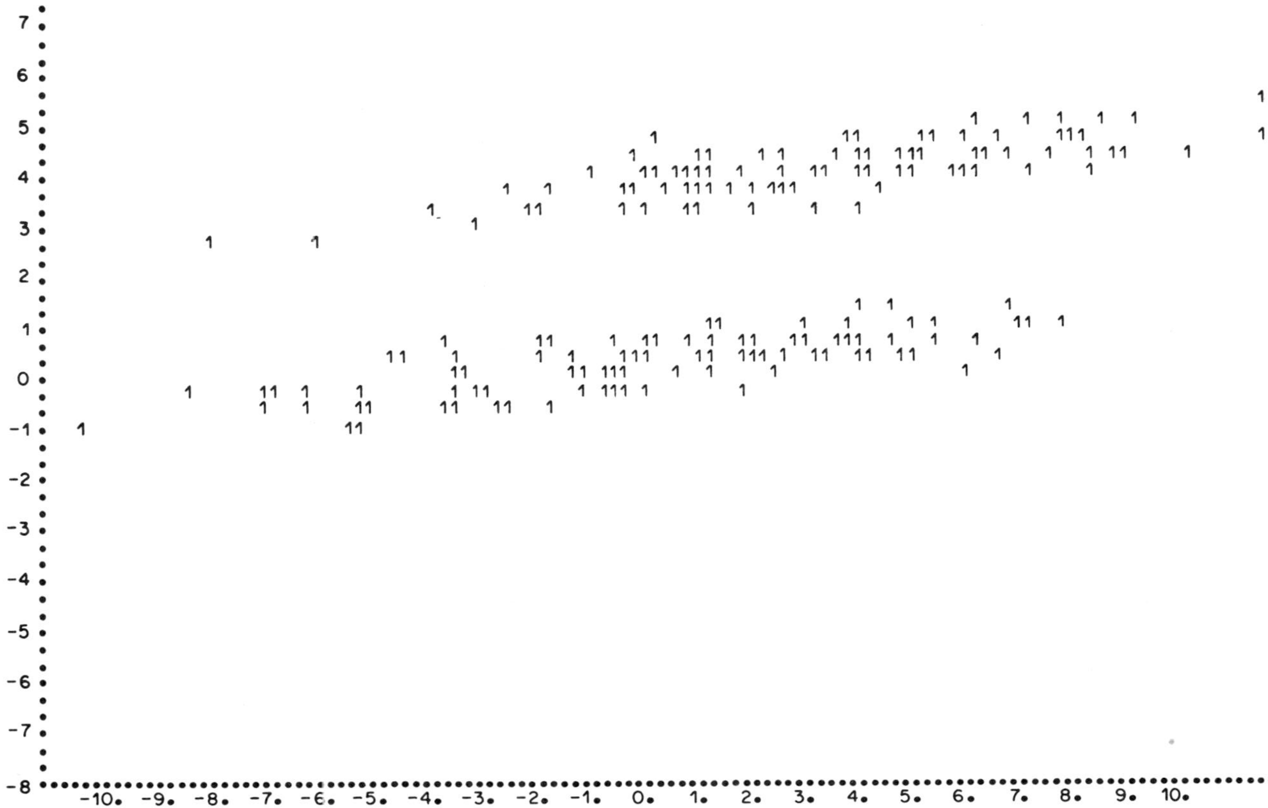

Figure 3.3. Data containing two well separated elliptical clusters.

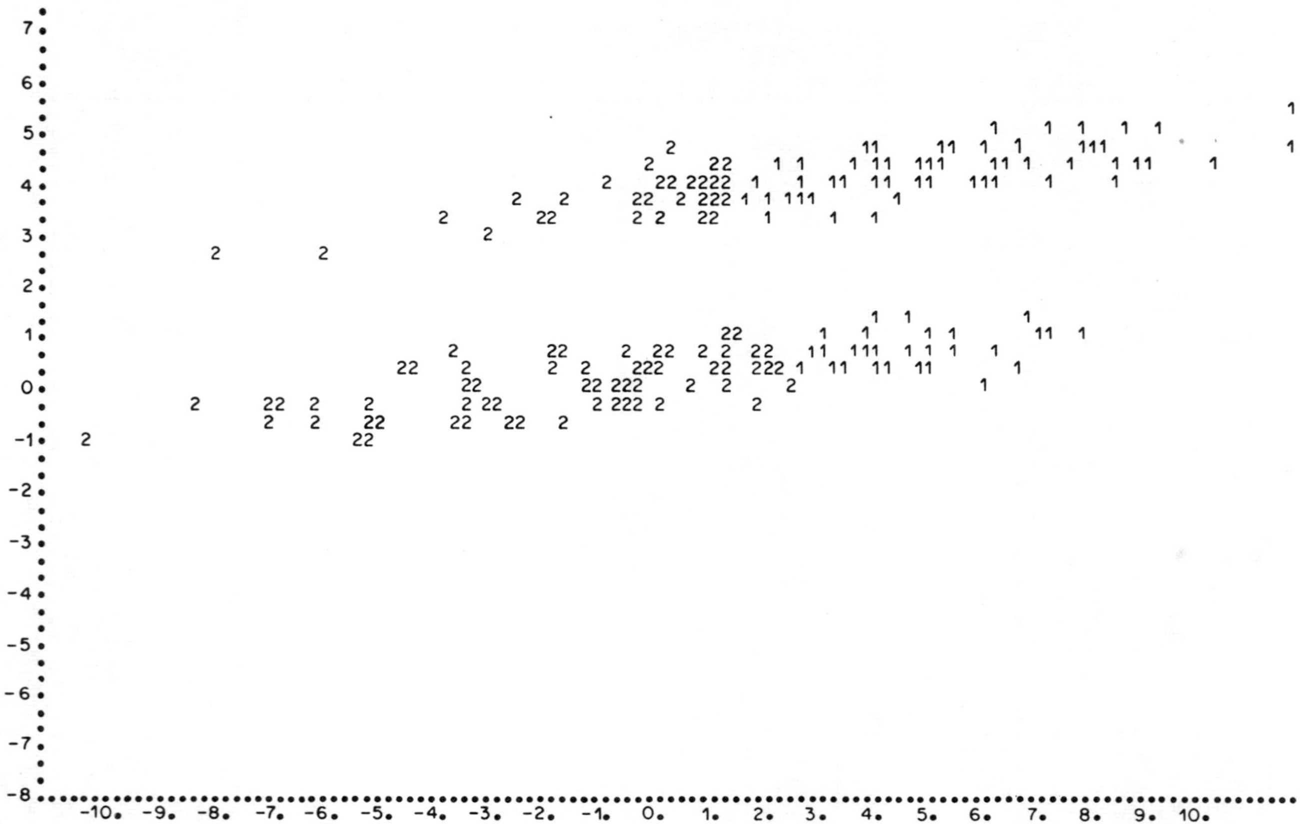

Figure 3.4. Two cluster solution given by within cluster sum of squares minimization clustering on the data shown in Figure 3.3.

3.2.4 Some Comments on Particular Clustering Techniques

A problem not mentioned in the previous section, but of considerable practical importance, is the plethora of clustering techniques available. The investigator interested in using clustering on his data is faced with an almost embarrassing number of possibilities. Ideally he should be presented with a list of good and bad techniques, and a recommendation of the best. Unfortunately this is a very difficult, if not impossible task, since it would seem that no one clustering method is 'best' in any sense for all situations. In this section therefore, we shall merely make a few comments about particular techniques which readers may find helpful.

The optimization techniques described in Section 3.2.2 usually require large amounts of computer time, and consequently are not suitable for use with large data sets (say with n of the order of one thousand and p around fifty). They do have an advantage over hierarchial techniques in containing a provision for reallocation of individuals who may have been poorly classified at some early stage in the procedure. Perhaps the most useful clustering criterion of those mentioned in section five is minimization of $|\mathbf{W}|$, the advantages of which are discussed in Marriot (1971) and Everitt (1974).

Hierarchial techniques require, in general, less computing time than optimization techniques and are, consequently, feasible for use with large data sets. A major difficulty with these techniques lies in the choice of one method from the many available and in the choice of which similarity or distance measure to use. Jardine and Sibson (1968, 1971), object to all hierarchial agglomerative techniques except single linkage clustering, on mathematical grounds. Their arguments are too complex for inclusion here, but they have not found universal acceptance since William *et al.* (1971) question their conclusion, and have found that techniques objected to by Jardine and Sibson on mathematical grounds have, nevertheless been far more helpful in providing useful information for the investigator than the mathematically acceptable single linkage clustering. Since in practice these techniques are used essentially for data-simplification and data description, it seems more reasonable to adopt this pragmatic approach than to restrict investigators *solely* to the use of single linkage clustering. The question as to which similarity or distance measure to use is one to which there can be no absolute answer. Sneath and Sokal (1973) discuss the problem in some detail. It should of course, be remembered that clustering techniques which operate on similarity or distance matrices may be thought of as attempting to summarize the information on the relationships between individuals given in the similarities or distances so that these relationships can be easily comprehended or communicated. Obviously the output of the clustering technique (the number of clusters and the cluster membership of individuals) will only be as meaningful as the input similarities and distances.

The cluster analysis technique of Wolfe based on the fitting of mixtures of multivariate normal distributions has a special appeal firstly because of its distributional model basis, and secondly because it involves no arbitrary choice of similarity or distance measure. A further advantage is the sequence of likelihood ratio tests for number of clusters which attends the method. (see below). The main disadvantages of this method are the large amount of computer time needed for moderately sized sets of data, and the problem of local maxima of the likelihood function. To overcome this latter problem several runs using different initial estimates of the parameters (that is the mixing proportions, the cluster means, and the cluster covariance matrices), are desirable. All other clustering techniques are potential sources of initial estimates of these parameters. An unknown factor with this method is how it is affected by departures from its distributional assumptions.

Likelihood Ratio Test for Number of Clusters

To test the hypothesis of say r' clusters against that of say r we may use the statistic $-2 \log \lambda$ where λ is the ratio of the likelihood of the observations under the opposing hypotheses. Wilks (1938) showed that under certain regularity conditions this statistic is asymptotically distributed as χ^2 with degrees of freedom equal to the difference in the number of parameters of the two hypotheses. Wolfe's clustering technique uses a slightly modified likelihood ratio test. The statistic used is

$$-\frac{2}{n}(n - 1 - p - r'/_2)\log \lambda,$$

and this is referred to a χ^2 distribution with $2p(r' - r)$d.f. This formula was arrived at by a Monte Carlo investigation (see Wolfe, 1971).

3.2.5 Evaluating Solutions

Once having obtained a set of clusters from some particular clustering technique the investigator is faced with the difficult problem of evaluating whether they represent, in some way, a convenient summary of the data, or whether they are of lasting importance in the particluar area under investigation in the sense of having some *predictive ability*, with respect to variables other than those used in their construction. For example, a new cluster analysis derived typology of mental illness would only be judged useful if it had some relevance for treatment. Again the classification of humans into males and females is based essentially on *one* particular fact of anatomy. It is however a very useful classification since many other variables differ in the two groups. On the contrary a classification of persons into those with hair on their forearms between $^3/_{16}''$ and $^1/_4''$ long and those without, though it may have some particular use, is certainly of no general use, for imputing membership in the former class to a person conveys information on this one property alone, there being, almost certainly, no other known properties which divide up a set of people in a similar manner. Consequently, investigators purporting to show that the clusters they obtain are of some importance, should pay considerable attention to the comparison of these clusters on sets of variables not used in the original analysis. It would, of course, also be of importance to replicate the solution on other data sets, as a further indication that the clusters were not simple artefacts; it should be remembered that the majority of clustering techniques will always produce clusters of *some* variety.

There are various procedures which may be helpful in evaluating solutions, and two of these are now described.

(1) Graphical Techniques

In many cases graphical representation may be extremely useful for displaying the results of a cluster analysis and giving insight into the merits of the clusters obtained. Graphic tools to display clustering should let one 'see' whether the grouping of the data has brought out any latent structure. Equally, if real structure of the cluster type is lacking in the data the graphic display of clustering results should bring this out. The most common graphical techniques for displaying the results of a cluster analysis are principal components plots and canonical variate plots. A more recent graphical technique which may be used when a similarity or distance matrix is involved is described by Ling (1972). This technique, known as SHADE, applied to the similarity matrix before clustering and to the rearranged matrix after clustering, shows at a glance whether clustering brought forth any distinctive clusters.

(2) Testing on Further Sets of Data

Perhaps the main method of overcoming subjectivity of interpretation in multivariate analysis in general and cluster analysis in particular is by replication that is by collecting further samples and repeating the analysis. The analysis of the first sample may possibly be used to generate hypotheses, which may then be tested on later samples. Of course, if one's sample is large enough such replication could be achieved by dividing the sample randomly into two and analysing each half separately. What should be avoided is using one sample to both generate and test the hypothesis. As remarked elsewhere (Williams and Dale, 1965), 'generation of the hypothesis cannot be used as its own evidence.'

3.3 EXAMPLES

3.3.1 Some Examples of the Application of Clustering Techniques

The number of papers describing applications of clustering techniques is considerable; in this section some of these will be briefly described. In the next section an application of clustering to some data on long term prisoners is described in detail.

An interesting application for cluster analysis in the field of market research is suggested by Morrison (1967). A large number of cities are available that could be used as test markets, but due to economic factors testing must be restricted to a small number of these cities. If the cities could be clustered into a small number of groups such that cities within a group were very similar to each other, then one city from each group could be used as a test market. Green *et al.* (1967) use this approach clustering a number of cities on the basis of variables such as city size, per capita income, etc.

There have been several reported applications of clustering in psychiatry. In this area the existing diagnostic categories have been extensively criticized, and many investigations, for example, Paykel (1971) and Everitt *et al.* (1971) have attempted to use cluster analysis techniques to find a more acceptable and more useful typology. Such investigations have been only partially successful although in a later paper (Paykel, 1972), the first of these authors shows that the typology he derived using cluster analysis was far more successful than the usual clinical diagnostic categories in predicting outcome when treatment with a particular therapeutic drug was considered.

Brennan (1971) described the use of clustering in a study of academic attainment amongst students. He forms groups which, in some sense, best predict a measure of academic attainment. These groups are then examined with respect to their members' ability to concentrate, their motivation, their aesthetic, social and religious values and other variables such as extraversion, and aptitude. The results give considerably insight into the relationship between academic attainment and social and motivational variables.

Entwistle and Bennett (1974) use cluster analysis on data describing psychological and educational characteristics of students. A large sample was available, and so this was divided randomly into two and separate analyses performed on each half. The clusters obtained from each analysis were very similar giving some indication of the stability of the groups. The student 'types' were also found to be associated with differing levels of academic performance, showing therefore that the 'types' have some useful predictive properties.

Recently there has been some interest in the use of cluster analysis techniques for stratification in sample surveys. For example, Golder and Yeomans (1973) use a particular technique of cluster analysis for stratification, in an investigation concerning the use of seat belts by motorists in Birmingham. Thirty nine wards in the city were clustered by a minimum variance type cluster method on the basis of thirteen socio-economic variables. The resulting five clusters of wards were then used as the basis of the sampling design of the investigation. Some other interesting examples of this use of cluster analysis are given in Dahmström and Hagnell (1974). Holgerson (1975) investigates the potential of different methods of cluster analysis as stratification 'tools' using as a criterion of effectiveness their ability to increase the efficiency of estimation.

Other applications of clustering techniques are described in McQuitty (1962), Wilkins and MacNaughton-Smith (1964) and Solomon (1971).

3.3.2 Numerical Example

In this section the results of applying a particular technique of cluster analysis to a set of data concerning long term prisoners will be discussed, in an attempt to illustrate more clearly how clustering may be useful.

(1) *The Data*

The data used in this example were kindly supplied by Dr. J. Gunn, Senior Lecturer, Institute of Psychiatry, London.

The data consisted of a set of eight scores for each of eighty-six long term prisoners in Wandsworth Prison, London. The eight scores were derived from the Buss-Durkee hostility questionnaire (Buss and Durkee, 1957), which consists of seventy-five questions having 'true (1) – false (0)' answers. Some typical questions from this document are shown in Table 3.1. The eight scores used in this investigation are subtotals of different sets of the original seventy-five answers; the scores measure various aspects of hostility, these being as follows:

1. Assault
2. Indirect Hostility
3. Irritability
4. Negativism
5. Resentment
6. Suspicion
7. Verbal Hostility
8. Guilt

Table 3.1. Some typical questions from the Buss–Durkee questionnaire

I seldom strike back, even if someone hits me first.	TRUE FALSE
I know that people tend to talk about me behind my back	TRUE FALSE
I demand that people respect my rights	TRUE FALSE

(2) *Clustering Technique Used*

The particular technique used to analyse this data was that described in Section 3.2.2, namely Wolfe's method of fitting mixtures of multivariate normal distributions. This particular technique was chosen because of its proven usefulness in other applications, and its availability in the form of a well documented computer program (see Appendix).

(3) *Results*

The three cluster solution obtained from Wolfe's NORMAP program (see Appendix) proved the easiest to interpret and therefore this will be discussed here.

Table 3.2 shows the within cluster means for the eight scores and the number of individuals in each cluster.

The first cluster showed low mean scores on all eight variables, indicating rather low hostility. This could perhaps be termed an essentially 'non-aggressive' group who on the whole, are likely to be less trouble for prison authorities. The other two groups show more hostility and could perhaps both be termed 'aggressive' groups. However, they differ in the manner in which their aggression is displayed. Group 2

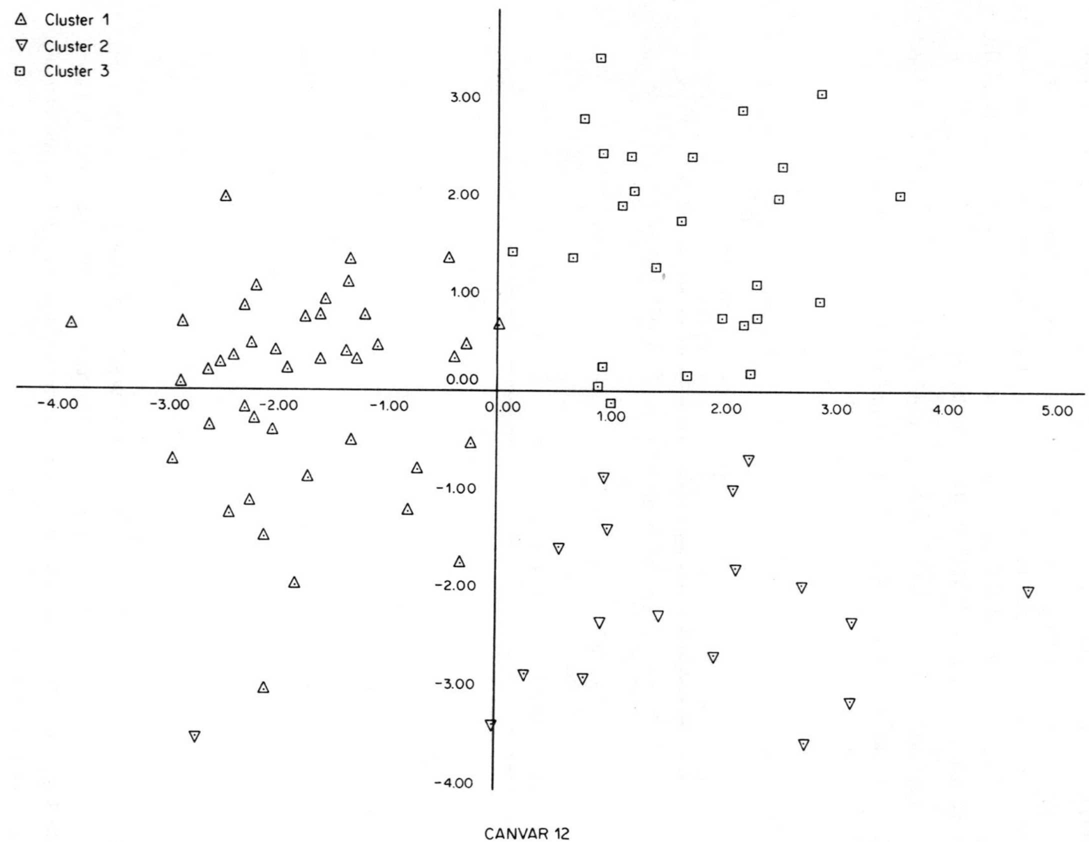

Figure 3.5. Canonical variate plot of three group solution produced by Wolfe's Normap clustering method on data concerning long term prisoners.

Table 3.2. Means on the eight variables for each of the three clusters derived from Wolfe's clustering technique

	No. in cluster	Variable 1	2	3	4	5	6	7	8
Cluster 1	43	5.5	3.5	5.0	2.9	2.9	3.3	6.4	3.9
Cluster 2	17	5.8	6.6	7.2	2.7	4.8	5.6	9.5	6.2
Cluster 3	26	6.4	3.1	4.5	3.0	4.6	5.1	7.0	5.3

seem more likely to show verbal hostility than actual assault; in group 3 hostility is likely to manifest itself more violently.

A graphical representation of these three clusters is given in Figure 3.5, which shows the individuals plotted in the space of the two canonical variates. (For a description of canonical variate analysis see Maxwell, 1968). The weights of these two variates are shown in Table 3.3.

The first canonical variate separates the first cluster from the other two. The variable having the highest weight in this variate is number five which is resentment, on which the two 'aggressive' clusters score similar to each other and higher than the 'non-aggressive' group. The second canonical variate separates the second two clusters. In this variate the variable which is most important is number two, that is, Indirect Hostility, and here the second cluster scores higher than the first.

In any 'in depth' study of this data the individuals comprising the clusters would be examined further, and the clusters compared with respect to other relevant variables such as nature of crime, length of imprisonment etc., in an effort to gain more insight into their potential usefulness as the basis for future studies, and their relevance for prison administration.

Table 3.3. Canonical variate weights

	Variate 1	Variate 2
1	−0.12	0.22
2	−0.02	−0.68
3	−0.41	−0.14
4	0.16	0.59
5	1.08	0.20
6	−0.04	−0.05
7	0.29	−0.22
8	−0.10	−0.06

3.4 SUMMARY

The term cluster analysis is used to describe a loosely structured body of *ad hoc* algorithms, which are useful for data exploration in many situations. The number of such techniques available is very large as is the number of problems in applying them, consequently care is needed in their use. They should be used in association with other multivariate analysis methods such as principal components and canonical variate analysis. In this way greater insight into the structure of the data will be obtained. The practice of simply applying one particular technique to a set of data, and accepting the solution produced is not to be recommended. Clusters should be evaluated in the light of results from other techniques, and with respect to their predictive ability, and interpretability in the context of particular area being investigated.

Like other methods of multivariate analysis, clustering is not a panacea for poor data, or a substitute for clear thinking. It is essentially a tool for data analysis which some investigators may find helpful. Like other tools for data analysis it must be used in a sensible manner if the investigator requires useful results. If not used with care the solutions given may be misleading and of little lasting value.

3.5 APPENDIX

In this appendix details are given of some of the packages of programs available for various methods of cluster analysis.

(i) Clustan 1B

This package of clustering programs is one of the most extensive available. Included in it are programs for single linkage clustering, Ward's method of clustering, Association analysis, Centroid clustering and many others.

Other options included choice of one of several similarity coefficients, obtaining dendrogram plots, including microfilm plots, obtaining principal components and using the principal component scores as input to the clustering techniques.

The package is very well documented and is relatively simple to use. Details may be obtained by writing to the General Sales Manager, London University Computing Services Limited, 39 Gordon Square, London WC1.

(ii) Friedman and Rubin Package

This package implements the methods of clustering described by Rubin (1967), and Friedman and Rubin (1967). It is available through the IBM share system for an IBM 360 machine and is almost certainly implemented at most 360 installations.

The package is well documented and simple to use. Facilities are provided for using random starting partitions in the optimization procedure. Use of this facility may give the user more confidence in the solutions, but it can be very time consuming.

(iii) Micka: A Fortran IV Iterative K-means Cluster Analysis Program

This program implements the same clustering techniques as the Friedman – Rubin package. It is, however, less comprehensive, having no facilities for generating different starting configurations, or for allowing these to be user supplied. It does have an advantage over the Friedman–Rubin package, since it could be implemented on any machine with a Fortran complier. It also appears to be faster than the Friedman–Rubin program.

A listing of this program is available from D. J. McRae, CTB/McGraw-Hill, Del Monte Research Park, Monterey, California.

(iv) Normap/Normix

This program fits mixtures of multivariate normal distributions by means of maximum likelihood methods. It includes facilities for reading in initial parameter estimates.

A well documented IBM version of this program is available, and details may be obtained from Dr. J. Wolfe, U.S. Naval Personnel Research Laboratory, San Diego, California.

REFERENCES

Andrews, D. F. (1972), 'Plots of high dimensional data', *Biometrics,* **28**, 125–136.

Ball, G. H. and Hall, D. J. (1967), 'A clustering technique for summarizing multivariate data', *Behaviour Sci.*, **12**, 153–155.

Ball, G. H. (1971), *Classification Analysis*, Stanford Research Institute SRI Project 5533.

Buss, A. H. and Durkee, A. (1957), 'An inventory for assessing different kinds of hostility', *Journal of Consulting Psychology*, **21**, 343–340.

Cattell, R. B. and Coulter, M. A. (1966), 'Principles of behavioural taxonomy and the mathematical basis of the taxonome computer program', *Brit. J. Math. Statist. Psychol.*, **19**, 237–269.

Chang, C. L. and Lee, R. C. T. (1973), 'A heuristic relaxation method for non-linear mapping in cluster analysis', *IEEE Transactions on Systems, Man and Cybernetics*, **3**, 197–200.

Chernoff, H. (1973), 'Using faces to represent points in *k*-dimensional space graphically', *J. Am, Statist. Assoc.*, **68**, 361–368.

Cormack, R. M. (1971), 'A review of classification', *Journal of the Royal Statis. Soc. Series A*, **134**, 321–367.

Dahmström, P. and Hagnell, M. (1974), 'The Formation of strata using cluster analysis', *Research Report No. 4*, Department of Statistics, Lund, Sweden.

Edwards, A. W. F. and Cavalli-Sforza, L. (1965), 'A method for cluster analysis', *Biometrics*, **21**, 362–375.

Entwistle, N. J. and Bennett, S. N. (1974), 'Cluster analysis in education research', Abstract in *Multivariate Behavioural Research*, **9**, 498.

Everitt, B. S. (1974), *Cluster Analysis*, London: Heinemann Press.

Everitt, B. S., Gourlay, A. J. and Kendell, R. E. (1971), 'An attempt at validation of traditional psychiatric syndromes by cluster analysis', *Brit J. of Psychiatry,* **119**, 299–412.

Everitt, B. S. and Nicholls, P. G. (1975), 'Visual methods for representing multivariate data', *The Statistician,* **24**, 37–49.

Fortier, J. J. and Solomon, H. (1966), 'Clustering procedures'. In *Proc. Symp. Multiv. Analysis, Dayton Ohio*, (P. R. Krishnaiah ed.), pp. 493–506, New York: Academic Press.

Friedman, H. P. and Rubin, J. (1967), 'On some invariant criteria for grouping data', *J.. Am. Statis. Ass.*, **62**, 1159–1178.

Golder, P. A. and Yeomans, K. A. (1973), 'The use of cluster analysis for stratification', *Applied Statistics*, **22**, 213–219.

Gower, J. C. (1967), 'A comparison of some methods of cluster analysis', *Biometrics,* **23**, 623–628.

Gower, J. C. (1971), 'A general coefficient of similarity and some of its properties', *Biometrics*, **27**, 857–872.

Green, P. E., Frank, R. E. and Robinson, P. J. (1967), 'Cluster analysis in test market selection', *Management Science*, **13**, 387–400.

Holgerson, M. (1975), 'Multivariate stratification with the use of cluster analysis', *Research Report, Department of Statistics*, Sweden: University of Uppsala.

Jardine, N. and Sibson, R. (1968), 'The construction of hierarchic and non-hierarchic classifications', *Comp. J.*, **11**, 117–184.

Jardine, N. and Sibson, R. (1971), *Mathematical Taxonomy*, London: Wiley.

Kendall, M. G. (1966), 'Discrimination and classification', In *Multivariate Analysis*. (P. R. Krishnaiah ed.) New York: Academic Press.

Kruskal, J. B. (1964), 'Multi-dimensional scaling by optimizing goodness of fit to a non-metric hypothesis', *Psychometrika*, **29**, 1–27.

Lambert, J. M. and Williams, W. T. (1966), 'Multivariate methods in plant ecology, IV', *J. Ecol.*, **54**, 635–664.

Ling, R. F. (1972), 'On the theory and construction of *k*-clusters', *Computer Journal*, **15**, 326–332.

Ling, R. F. (1973), 'A computer generated aid for cluster analysis', *Communications of the ACM* **16**, 355–361.

MacNaughton-Smith, P. (1965), *Some Statistical and Other Numerical Techniques for Classifying Individuals*, Home Office Research Unit Report No. 6. London H. M. S. O.

MacNaughton-Smith, P., Williams, W. T., Dale, M. B. and Mockett, L. G. (1964), 'Dissimilarity analysis', *Nature*, **202**, 1034–1035.

MacQueen, J. (1966), 'Some methods for classification and analysis of multivariate observations', *Proc. 5th Berkeley Symp.* **1** 281–297.

Marriot, F. H. C. (1971), 'Practical problems in a method of cluster analysis', *Biometrics*, **27**, 501–514.

Maxwell, A. E. (1968), 'Multivariate analysis: Some problems in the discrimination and classification of pyschiatric patients', *The 5th International Scientific Meeting of the International Epidemiological Association*.

Morrison, D. G. (1967), 'Measurement problems in cluster analysis', *Management Science*, **13**, 775–780.

McQuitty, L. L. (1962), 'Multiple hierarchial classification of institutions and persons with reference to union-management relations and psychological well-being', *Educ. Psychol. Measure*, **23**, 55–61.

Paykel, E. S. (1971), 'Classification of depressed patients: A cluster analysis derived grouping', *British Journal of Psychiatry*, **118**, 275–282.

Paykel, E. S. (1972), 'Depressive typologies and response to amitriptyline', *Brit. J. Psychiatry*, **120**, 147–156.

Sammon, J. W. (1969), 'A nonlinear mapping for data structure analysis', *IEEE Trans Computers*, C18, 401–409.

Singleton, R. C. and Kautz, W. (1965), *Minimum Square Error Clustering Algorithm*. Stanford: Research Institute.

Sneath, P. H. A. and Sokal, R. R. (1973), *Principles of Numerical Taxonomy*. London: Freeman.

Solomon, H. (1971), 'Numerical taxonomy' in *Mathematics in the Archaeological and Historical Sciences*. Ed. F. R. Hodson, D. G.. Kendall and P. A. Tauty, Edinburgh: University Press.

Wilkins and MacNaughton-Smith (1964), 'New prediction and classification methods in criminology', *J. Res. in Crime and Delinquency*, 1.

Wilks, S. S. (1938), 'The large sample distribution of the likelihood ratio for testing composite hypothesis', *Ann. Math. Stat.*, **9**, 60–62.

Williams, W. T. and Dale, M. B. (1965), 'Fundamental problems in numerical taxonomy', In *Advances in Botanical Research*. Ed. R.D. Preston. London: Academic Press Vol. 2.

Williams, W. T., Lance, G. N., Dale, M. B, and Clifford, H. T. (1971), 'Controversy concerning the criteria for taxonometric strategies', *Computer Journal*, **14**, 162–165.

Wolfe, J. H. (1967), *NORMIX: Computational Methods for estimating the parameters of Multivariate Normal Mixtures of Distributions*., Research Memorandum SRM 68–2. San Diego: U.S. Navy Personnel Research Activity.

Wolfe, J. H. (1969), *Pattern Clustering by Multivariate Mixture Analysis*, Research Memorandum SRM 69–17, San Diego: U.S. Navy Personnel Research Activity.

Wolfe, J. H. (1970), 'Pattern clustering by multivariate mixture analysis', *Multiv. Behavioural Research*, **5**, 329–350.

Wolfe, J. H. (1971), 'A Monte Carlo Study of the sampling distribution of the likelihood ratio for mixtures of multinormal distribution', *Naval Personnel and Training Research Laboratory*. (San Diego, California 92152). Technical Bulletin STB 72–2.

Chapter 4

Principal Component and Factor Analysis

C. C. Taylor

4.1 Introduction

This chapter deals with two techniques of analysis: principal component analysis (PCA) and common factor analysis (CFA). PCA has a long history as a statistical technique for analysing multivariate data. It is usually attributed to Hotelling (1933) although earlier descriptions of the method may be found in the literature. It is a well-defined and frequently described procedure which appears in all standard texts on multivariate analysis; recent, detailed descriptions are easy to find, for instance Kendall and Stuart (1961). By contrast factor analysis, as the term is currently used, is comprised of a set of widely differing techniques that have arisen in a piecemeal fashion since the beginning of the century when Spearman (1904) first introduced the conceptual framework. During its development some of its procedures have sometimes lacked clear definition and it seems there has been difficulty at times in separating its objectives from the numerical algorithms that are used. Furthermore it has often not been distinguished adequately from PCA in its application, a feature which persists in current practice. It is now generally accepted that the definitive statistical formulation is that provided by Lawley and Maxwell (1971) and to distinguish this from the generic 'factor analysis' it is here referred to as common factor analysis, as in Kaiser (1970). The principal numerical algorithm used in current analyses, although not the only one, is that of Joreskog (1967). Standard texts on factor analysis, for instance Harman (1960) or more recently Gorsuch (1974), usually give detailed descriptions of both PCA and CFA and in addition discuss a varied assortment of associated techniques. The more mathematical text of Lawley and Maxwell deals primarily with CFA.

PCA and CFA are covariance/correlation analyses that operate within the framework of linear models. As such, they are techniques which require data measures on a full interval scale. They may be loosely described as attempting to condense a set of observed variables into a smaller number of constructed, conceptual variables – the components, or factors – that have not themselves been directly measured. (Technically, in PCA one may use a number smaller than or equal to the number of variables, but the extreme case is almost never encountered in practice.) The observed relationships between the measured variables are viewed as being due to a structuring of the data, which these conceptual variables can be used to describe. PCA and CFA differ in the principles they use to construct the unmeasured variables, but the feature common to all factor and component methods is that they are used to identify and describe an underlying structure beneath a set of multivariate data.

Within this loose description there are various types of particular interpretation of the analysis that can be made, depending upon the precise use to which it is put. For example, these techniques are most frequently used in psychology, where one might find a battery of tests being given to a group of subjects. Each test question could be considered as seeking to measure some underlying characteristic such as intelligence, or at least contributing towards its measurement. The multivariate data set, that is, the test battery, is thus assumed to be generated by a single underlying variable (intelligence) which has been subjected to various distortions and extraneous influences. This structure is what the analysis should identify. Alternatively, one may consider the analysis of a set of data where the structure is not explicitly causal, but exists merely in that there is a stable pattern of covariation. For instance, a survey might produce a data set comprising of a number of different measures of economic well-being and the aim of the analysis might be to derive a summary index of economic status. Clearly here the underlying characteristic of status cannot be thought to generate the economic measures in the same way that intelligence might generate the test results. However, component or factor methods could be used in absence of any detailed economic theorizing to provide an index of status.

The determining of a *single* underlying characteristic is not the crux of the analysis. It is only required that there be fewer underlying concepts than there are measurements. At one time whether intelligence was intrinsically a one-dimensional or multi-dimensional phenomenon was a much-debated question (Spearman, 1904; Cattell, 1952) and factor analytic methods were developed in the attempt to shed light on this and other such problems. In the example of a survey that requires an index of economic well-being, it is easy to see that one may allow more than one conceptual dimension to well-being and so require several indices, each simultaneously measuring interrelated aspects of the underlying concept. In survey work PCA and CFA methods lend themselves most easily to attitude questionnaire analyses and the like. Here it could be presumed that there exist general, compound attitudes that underpin the responses to the particular questions asked. An obvious example might be political alignment in a simple rightist-leftist dimension, which could be presumed to give some structure to the answers to a set of questions dealing with current political issues. In surveys of more specialist populations, questionnaires dealing with mood or personality assessment are analysed in a similar fashion. Here the questionnaires clearly use a set of 'indicator' items that are intended jointly to elicit a more fundamental feature of the respondent's character or behaviour. Other sorts of analyses using dimensions which are less clearly defined also occur. Moser and Scott (1951) in a survey of British towns seek to characterize various social and demographic aspects by reducing them to a low-dimensional structure determined by a few derived variables. In the biological sciences examples are found (e.g., Matsakis, 1957) of smaller scale studies which seek to describe physical differences between various related species. Physical measurements of body-parts are used to develop indices of shape or size by which to compare one species with another.

These and the examples to be quoted subsequently show only a small selection of possible fields of application and types of use for the analyses. In essence all these problems are the same. The measured variables, as in any multivariate analysis, must in some sense jointly describe a domain of study, or more likely, are used to provide an *ad hoc* definition of some such domain. Within this domain stable relationships between the variables exist, which create a simple, underlying dimensionality to the problem; the variables themselves are considered as multifarious measurements imposed on or related to these few dimensions. The aim of the analyses is to show the relationship between the observed variables and the identified sub-structure.

Thus in its mathematical formulation we have a set of unobserved measurements, denoted **F** (which are the components or factors), which relate to the observed data, denoted **Y**, through a loadings matrix **L**. The loadings matrix **L** gives fundamental information about which variable relates to which factor and how strongly. These are effectively *path coefficients* to each variable from the factors. The loadings pattern and the factors together produce the structure in the data, and allowing for errors of fit of one sort or another (denoted **E**) we have a fundamental decomposition of the data matrix

$$\mathbf{Y} = \mathbf{LF} + \mathbf{E}$$

The elements of this decomposition have different properties depending upon which analysis, PCA or CFA, is used and are discussed in detail in the subsequent sections. In general the analyses concentrate on the structural coefficients in **L** and on the components or factors **F**. These in conjunction effect a reconstruction of the data matrix, differing from the original only by the residual errors **E**. In some instances, it is this reconstruction $\mathbf{Y}^*$ that is of more importance than the factors **F**. It may be considered as an approximation to the observed data **Y**, but one which has a simple low-dimensional structure to it expressible in the form

$$\mathbf{Y}^* = \mathbf{LF} \quad \text{where} \quad \mathbf{Y} = \mathbf{Y}^* + \mathbf{E}$$

The usual assumption in using this approach is that it is the unimportant bits of the data that are in this way discarded, to leave a 'better' data matrix $\mathbf{Y}^*$.

The overall approach to the subject matter of this chapter has been determined primarily by two considerations. Firstly, it is argued that in practical problems it is not usually clear whether component or factor analysis should be used: arguments can be made for or against either technique. Therefore an attempt has been made to avoid polarizing the descriptions of these two closely related techniques, but instead to point up their similarities and thus show to what extent the one may substitute for the other. It now seems almost compulsory, when writing about PCA and CFA, to pay lip service at some stage to the viewpoint that CFA generates testable hypotheses, whilst a complete PCA is a simple data transformation; and as such there can be no interchangeability. Let it suffice to say here that the present writer does not consider these two properties to characterize the vital distinction between the two analyses. If one might anticipate the arguments later in the chapter with two simple comments: in practice PCA is almost always successfully used in the truncated form described here, in a fashion similar to a factor analysis; again in practice CFA is often used, justifiably, without regard to its hypothesis testing potential. Hopefully these points will be made clear in subsequent sections.

Secondly, there is an attempt to avoid embedding the description of the techniques in the framework of a multivariate normal distribution. Although this approach would allow a close link-up with classical statistical theory it would not be so readily applicable to practical data analysis. Most data are not multinormally distributed, and it is as well to start with this in mind; to assume normality, and then provide lengthy qualifications to cover possible non-normality obscures the power of the simpler concepts of correlation, variance and least squared deviation which are the basis of these analyses. Such statistics are most natural in linear structures and these latter are unhesitatingly adopted here, on grounds that they will usually privide a robust first approximation for many problems. Thus, although there is a great deal of theoretical work at present on order invariant and even non-metric procedures for factor and component analysis, this is mentioned only in passing, at the end of the chapter.

The subject matter of the chapter is covered in four sections, of which the next deals with the description and interpretation of the loadings matrix **L**. The interpretation involves some general considerations in using linear combinations of variables and the simple geometric use of axes. This is dealt with through use of regression and partial correlation statistics, which provide a natural way of describing the properties of linear models. Section 4.3 deals with PCA and Section 4.4 contrasts the use of CFA; in these sections no further knowledge of statistical methods is assumed beyond that of correlation and regression.

*Mathematical Notation

Most of the mathematics necessary for the description of the techniques has been confined to asterisked sections and may be omitted by the reader.

Where algebraic notations is used, it is consistently as follows:

n	Number of observations
p	Number of variables
$\mathbf{Y}$	Data matrix, observations as columns ($p \times n$)
$\mathbf{S}$	Covariance matrix for $\mathbf{Y}$ ($p \times p$)
$\mathbf{C}$	Eigenvector matrix for $\mathbf{S}$ ($p \times p$)
$\mathbf{L}$	Loadings matrix ($p \times q$)
$\mathbf{F}$	Factor or component scores matrix ($q \times n$)
$\mathbf{E}$	Residual or error matrix ($p \times n$)
$\mathbf{V}$	Covariance matrix for $\mathbf{E}$ ($p \times p$)
λ_j	Eigenvalues of $\mathbf{S}$
θ_j	Eigenvalues of $\mathbf{V}^{-\frac{1}{2}}\ \mathbf{S}\ \mathbf{V}^{-\frac{1}{2}}$
s_j^2	Diagonal elements of $\mathbf{S}$
v_j^2	Diagonal elements of $\mathbf{V}$

Generally y_{ji}, $\mathbf{Y}_j$, $\mathbf{Y}^*$ are the elements, columns and approximation, respectively, for $\mathbf{Y}$; similarly for the other matrices.

These sections are intended only to sketch out the derivations that are sometimes needed, and to make more precise some of the verbal definitions and descriptions in the main body of the chapter. For detailed proofs the reader is referred to specialist texts such as Rao (1968), Kendall and Stuart (1966), and Lawley and Maxwell (1971).

4.2 AXIS ROTATION AND LINEAR BASIS

The analyses discussed above have been described as using the concept of an underlying structure which in some sense gives rise to the observed variables. It is sometimes useful to represent this in geometric terms and refer to this structure as a low-dimensional subspace within the overall space covered by the observed variables. In many analyses this subspace can usefully be regarded as providing an approximation to the overall set of variables, as the component or factor space. By constraining this space to a relatively low dimensionality one is building in simple structural relationships between the variables. In using this (linear) subspace to typify structure in the data we require only that certain unspecified linear relationships can be found amongst the variables; the generality that this formulation affords should adequately cover many practical situations. Thus it should be noted that the geometrical interpretation described in 4.2.1 merely a convenience and in no way critical; the analysis may be viewed simply in terms of linear combinations of variables if this is more suitable.

Almost all factor and component methods have two quite distinct stages: firstly, the identification of the approximating space, and secondly, the locating of co-ordinate axes within that space by which to describe it. (One exception is what Lawley and Maxwell (1971) call 'restricted model factor analysis' (see Section-4.4.4). To use a very simple analogy, the first stage of the analysis determines a terrain and the second draws a map to describe it. Locating and rotating axes in the space provides a way of exploring the different relationships among the variables and between the approximating space itself and the variables. The principles of axis rotation in, or put more accurately of choosing a *linear basis* for, the subspace which are used in the second stage of analysis are common to both PCA and CFA. For this reason they are described here before detailing the methods of PCA and CFA by which the subspace is chosen. In this description it is convenient to draw extensively on the concepts of correlation and regression analysis (see Chapter 2 in Volume 1), and especial use is made of their geometric representations. Factor analysis has developed its own vocabulary to deal with these concepts but it should be borne in mind that this is simply a renaming of the standard properties of linear models.

4.2.1 A Geometric Interpretation

The use of linear structures in conjunction with the concepts of variance and correlation leads naturally to analysing the data in terms of the ordinary concept

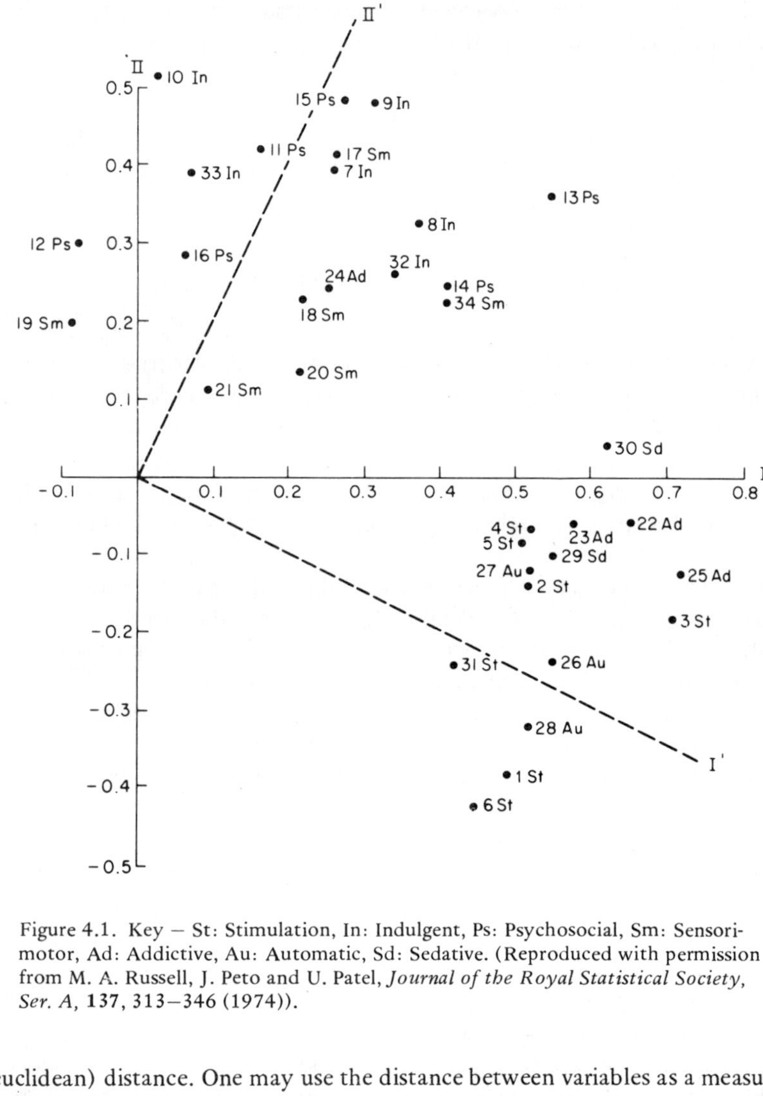

Figure 4.1. Key – St: Stimulation, In: Indulgent, Ps: Psychosocial, Sm: Sensorimotor, Ad: Addictive, Au: Automatic, Sd: Sedative. (Reproduced with permission from M. A. Russell, J. Peto and U. Patel, *Journal of the Royal Statistical Society, Ser. A,* **137**, 313–346 (1974)).

of (euclidean) distance. One may use the distance between variables as a measure of their dissimilarity in an obvious fashion. For example, Figure 4.1 shows diagramatically a two-dimensional subspace onto which have been projected several observed attitude variables. The analysis is discussed in more detail as Examples 1 and 3, below but here is used only to point out that the distances between points are a natural and obvious way to describe the relationships between the variables. The attitude variables, which were previously grouped into seven categories, as shown in Table 4.1 can be seen to fall into two loosely correlated groups: the association between the Stimulative, Addictive and Automatic items being quite close; and items from the remaining categories forming a group with perhaps a slightly lower general level of association.

The second important set of relationships is the level of association of each

Table 4.1. (Reproduced with permission from M. A. Russell, J. Peto and U. Patel, *Journal of the Royal Statistical Society, Ser. A*, **137**, 313–346 (1974))

	Questionnaire items
(1)	I like smoking while I am busy and working hard (St)
(2)	Smoking helps to keep me going when I'm tired (St)
(3)	Smoking helps me to think and concentrate (St)
(4)	Smoking cheers me up (St)
(5)	I get a definite lift and feel more alert when smoking (St)
(6)	I smoke more when I am rushed and have lots to do (St)
(7)	I want to smoke most when I am comfortable and relaxed (In)
(8)	After meals is the time I most enjoy smoking (In)
(9)	I like a cigarette best when I am having a quiet rest (In)
(10)	I usually only smoke when I can really sit back and enjoy it (In)
(11)	I smoke for the pleasure of offering and accepting cigarettes from other people (Ps)
(12)	I feel more mature and sophisticated when smoking (Ps)
(13)	It is easier to talk and get on with other people when smoking (Ps)
(14)	While smoking I feel more confident with other people (Ps)
(15)	I smoke much more when I am with other people (Ps)
(16)	I feel more attractive to the opposite sex when smoking (Ps)
(17)	Handling a cigarette is part of the enjoyment of smoking it (Sm)
(18)	I smoke for the pleasure of having something to put in my mouth (Sm)
(19)	Part of the enjoyment of smoking is watching the smoke as I blow it out (Sm)
(20)	Part of the enjoyment of smoking comes from the stops I take to light up (Sm)
(21)	I smoke because I like the smell so much (Sm)
(22)	When I have run out of cigarettes I find it almost unbearable until I can get them (Ad)
(23)	I am very much aware of the fact when I am not smoking (Ad)
(24)	Without a cigarette I don't know what to do with my hands (Ad)
(25)	I get a real gnawing hunger to smoke when I haven't smoked for a while (Ad)
(26)	I smoke automatically without even being aware of it (Au)
(27)	I light up a cigarette without realizing I still have one burning in the ashtray (Au)
(28)	I find myself smoking without remembering lighting up (Au)
(29)	I smoke more when I am worried about something (Sd)
(30)	I light up a cigarette when I feel angry about something (Sd)
(31)	I smoke in order to keep myself from slowing down (St)
(32)	One reason I smoke is because it tastes so good (In)
(33)	I only really enjoy smoking with a drink (In)
(34)	I find it a pleasure drawing the smoke into my lungs (Sm)

Key: St = Stimulation. Sd = Sedative.
Sm = Sensorimotor. Au = Automatic.
In = Indulgent. Ps = Psychosocial.
Ad = Addictive.

variable with the subspace onto which it has been projected. Clearly these distances cannot be shown directly on a diagram such as Figure 4.1 since they are, so to speak, distances from the plane of the page. It is therefore usual to quote these figures separately. However, in the case where variables have been effectively scaled to have equal variance, as is the most common situation, the distance from the approximating space varies inversely with the length of the line joining the origin to the projection point. Thus under these circumstances variables close to the origin are poorly associated with the sub-space, being at some distance from it; whereas variables distant from the origin are more closely associated with the subspace. The nature of these associations is what the analyst must primarily use in attempting to identify the sort of dimensions the subspace represents. This point is taken up in Example 1.

In the subspace it is possible to draw arbitrary axes to provide a linear basis, and in the present context these axes are called factors or components of the space. For example, in Figure 4.1 a point might be drawn anywhere out along axis I, say, and be treated as a new derived variable representing axis I (i.e. factor or component I) in its relationships with the other projected variables. It is an important feature of linear structures that these projected variables may be obtained by a transofrmation of *any* basis; or put geometrically, they may be described with reference to any set of axes. It is this property that is exploited by the techniques of axis rotation which are now discussed. Before doing so, however, it should be pointed out that there is no analytical reason why the basis should not be rotated to give obliquely angled

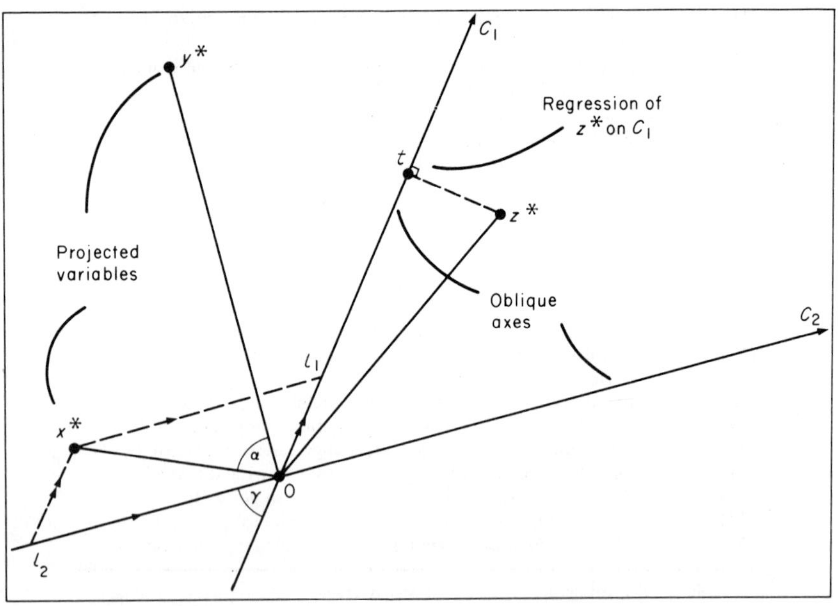

Figure 4.2. Geometric representation of variances and covariances:

x^*, y^*, z^*	points representing projected variables
C_1, C_2	axes representing components or factors (obliquely rotated)
l_1, l_2	loadings on axes 1 and 2 for variable x^*

axes; and a short digression may be useful to point out the implications of such a rotation.

Figure 4.2 shows a two-dimensional approximating space for three variables x, y, z. The resultant projection points x^*, y^*, z^* have been referred to oblique axes C_1, C_2 and origin O. The loadings (or coordinates) of x^*, for example, are shown as l_1, l_2 and their magnitude indicates how close x^* is to C_1 or C_2. (They are obtained by drawing the broken lines to the respective axis, parallel to the other). Thus with oblique axes, the loadings may still be used to describe the relative positions of x^*, y^*, and z^*. One feature to note therefore is that the projection of z^* onto the dimension C_1 at point t, which gives the regression of z^* on C_1, is not the loading of z^* on C_1. With orthogonal axes the two are identical, as we shall see subsequently. The angle α (or more precisely, $\cos \alpha$) indicates the correlation between x^* and y^*; and similarly $\cos \gamma$ indicates the correlation between the two axes or factors. The level of association between the approximating space and the variables, i.e. the distances of the variables, x, y, z from the space in which x^*, y^*, z^* lie, are clearly not representable in a diagram such as Figure 4.2 and must be quoted separately. It is usual to quote not the distance but rather the squared multiple correlation (SMC) of a variable with the subspace; and similarly to use not the distance between points x^* and y^*, for example, but rather their correlation $\cos \alpha$. That none of these correlations (between variables) are affected by the location of the axes is the central property that allows axis rotation. The rotations change only the correlations amongst the factors, and the correlations between the factors and the variables.

4.2.2 Properties of rotation

Returning now to the general purpose of axis rotation, it can be seen that when dealing with two dimensions it is easy to make a plot of the information, as in Figure 4.1. In order to cope with more than two dimensions, use is made of the freedom to rotate the axes in the space. Since the relationships between the variables, and the size of their SMC's with the approximating space are invariant under axis rotation, the axes may be drawn arbitrarily and the loadings on the axes be used as guides to the configuration of the variables. From Figure 4.1 it can be seen that relocation of the axes might make a considerable difference to the loadings patterns. These relationships, and in particular the SMC's, should give an indication to what sort of dimensions the approximating space spans. Hopefully it is the sort of structure anticipated when first selecting the variables in the data set; or at least one that can be subsequently interpreted in terms of the variables.

Techniques for locating the axes have been developed principally in factor analysis settings, where it is customary to present several further items of information to aid the final interpretation. Although the nomenclature used is different, the concepts involved here are simply those of correlation and regression. The usual items are as follows:

(i) Clearly the angle between the axes is a prerequisite – in Figure 4.2 this is marked γ – and its cosine is the correlation between the axes (factors); a right angle denotes zero correlation. Thus the matrix of correlations between factors is important in indicating the degree of separation between factors. It is sometimes (erroneously) suggested that if two factors are highly correlated, the analyst might 'make do with' one; but it should be obvious that, *ceteris paribus*, this correlation does not relate to the adequacy of a one- or two-dimensional representation of the data. This is a separate issue to which we return later.

(ii) The simple loadings matrix (pattern matrix) is in effect the matrix of partial regression coefficients of the variables on the axes (components), and has all its usual interpretations. Thus in general x^*, for example, may be taken as the multiple regression of x on dimensions C_1, C_2 with coefficients l_1, l_2. The length (Ox^*) squared is the variance of the regression. and taken in ratio to var(x) gives the SMC. Bearing in mind that var(x) basically measures the differences between individuals on variable x, the SMC shows the extent to which these differences lie in (i.e. are attributable to) the dimensions C_1, C_2. It is worth emphasizing that this particular measure does not depend upon the pattern of loadings in the chosen axis rotation.

(iii) The line (Ot) is the regression of the observed variable z (and also incidently of z^*) on dimension C_1. If the variance of the observed variable is standardized to unity, then the length of this line represents the simple correlation of the variable z with dimension C_1. The matrix of these correlations is called the structure matrix, and is usually as important as the pattern matrix in helping interpret the results of the analysis.

(iv) In addition, the matrix of partial correlation coefficients of the variables with the dimensional axes (known as the reference vector correlation matrix) is often used. Specifically, this gives the correlation of a variable with each dimension in turn, partialling on all the others. As such, it is simply a transformation of the pattern matrix in (ii) above.

It is not difficult to see that when the axes (factors) are orthogonal, all three matrixes (pattern, structure and reference vector correlation) are identical.

The general principles can be illustrated by looking ahead again to Example 1, Figure 4.1, and taking note of the two orthogonal axes I, II (solid lines). The group of fairly closely correlated variables are reasonably highly correlated with dimension I, whilst variable 13 has a moderate correlation with dimension II also. Some of the more loosely correlated set, for example 20 and 21, are only poorly represented by the two-dimensional space of the diagram. These remarks are taken up in Example 1. The point to be made is that in the more complicated situations the statistics of regression analysis can be used in conjunction with judiciously located axes to replace simple diagrammatic maps like figure 4.1.

The selection of criteria by which most conveniently to position the axes must be an arbitrary choice by the data analyst, in that they are not suggested by the statistical properties of the data. One of the oldest methods is to use the *principal axis* rotation, a method which is in effect an application of PCA. The first axis is selected in the direction of maximum variation across the points, that is, to maximize the (squared) size of the loadings on this axis; the second, in the direction of maximum variance but constrained to be orthogonal to the first; and so on. In practice, this is not often a useful positioning. For sets of data that display a general level of intercorrelation, this method tends to produce a 'general factor', i.e. an axis on which most variables load rather highly, with subsequent axes differentiating to a limited extent between small groups of variables. Almost invariably a clearer picture may be obtained by setting at least subsequent axes by different means.

Thurstone (1947) set out five principles which, from a purely factor analytic standpoint, he considered would define what he termed a 'simple structure'. These have proved difficult to cast into a mathematical form, and the subtleties of these principles have now been trimmed off to produce an approach which basically seeks to polarize the loadings to very high (positive and negative) or very low (zero) values. Criteria to achieve this have proliferated in mathematical psychology and carry the exotic names of that discipline. Carroll (1953) and Neuhaus and Wrigley (1954) independently proposed that *quartimax* criterion. Axes are rotated, maintaining orthogonality, to maximize the totalled variances of the squared loadings for each variable. This method has been found to frequently give a general factor, as does principal axis rotation. Kaiser (1958) suggested the *varimax* criterion, which removed this tendency by maximizing the totalled variance across the squared loadings for each factor instead of for each variable. Both these methods play down the significance of the variables that lie close to the origin, and so it is now usual to employ the 'Kaiser normalization' which weights the loadings to give each variable equal importance in the maximization criterion. For oblique axes, the *oblimin* criterion (Harman, 1960) provides a rotation that maximizes, averaging over all pairs of axes, a weighted average of the covariance and raw crossproduct moment of the squared loadings. By shifting emphasis on the weighted average a more highly oblique positioning may be achieved, at the discretion of the analyst. *Promax* (Hendrickson and White, 1964) is a computationally efficient method which adjusts axes from an initial orthogonal position by attempting to further polarize the loadings. Harris and Kaiser (1964) give a set of methods called *orthoblique* which are invariant under rescaling of the axes, (i.e. with respect to the variances of the factors; see Section 4.3).

A second, and possibly more useful, approach to the problem is found in Procrustes rotations. The axes are rotated until they match as nearly as possible (in at least squares sense) to a previously specified configuration; hence the data analyst is required to know what configuration to expect. Harley and Cattell (1962) give the oblique rotation procedure, and Schonemann (1966) discusses both oblique and orthogonal procedures. Because these techniques rotate to an hypothesized structure, they tend to be used in applications other than the exploratory studies for which the simple structure group was designed. The Procrustes methods are used most often in matching results from one study to results from a parallel study, especially in fields other than factor and component analysis.

A detailed comparison of rotation methods can be found in Gorsuch (1974), including some comments on their stability under repeated sampling. No rules can be made for picking a best method. It is, rather, a matter of finding the most informative method for the data being analysed.

4.2.3 Example 1

The following example is taken from a study of attitudes toward smoking in Russell, Peto and Patel (1974). The items on the (shortened) questionnaire are given in detail in Table 4.1 above. A two-dimensional approximating space was fitted to the 34 attitudes that were measured. Figure 4.1 above plots the projections of these variables onto the two dimensions. Principal axes (I, II) are shown, along with the varimax rotated axes (I′ II′). It can be seen that the items fall into two broad classes, the first consisting largely of addictive and stimulative questions – 'to keep from slowing down', 'gnawing hunger to smoke' – and the second, a variety typified by psychosocial attitudes – 'feel more confident'. The first principal axis shows the usual 'averaging' effect in that the majority of variables load medium to high on it. The second principal axis clearly separates the two classes. The varimax rotation, by contrast, groups the first set around axis I′, with near-zero loadings on axis II′; and the second set *vice versa*.

The authors discuss the possibility of interpreting the first principal axis, I, as measuring pharmacological dependence on nicotine. Thus the items in the first set would be presumed highly predictive of pharmacological dependence; and those in the second set, whilst mainly indicative of other motives for smoking, are also weakly predictive of nicotine dependence. The authors point out, by contrast, if the varimax axis I′ is presumed to be the pharmacological dimension, the items in the second set are virtually independent of nicotine dependence. It can be seen that the choice of rotation technique can critically affect the final interpretation of an analysis.

All the items were initially scaled to have unit variance, so their (squared) distance from the origin represents their SMC with the approximating space. It can be seen that most of those in the 'addictive' group have a relatively strong association with this space, and are therefore well-presented by the two-dimensional analysis. Some of the 'psychosocial' group, however, are not so closely correlated, and have a considerable variability in dimensions other than the two shown. Thus one could not with confidence use these variables in interpreting the dimensions shown.

As a final point, it should be noted that although both the rotations are to orthogonal axes, an oblique positioning obtained from one of the simple structure groups (e.g. promax) would not give anything very different from the varimax solution. Whether the possible improvement in simplicity of loadings pattern is worth the expense of introducing correlation between the factors is a moot point.

4.2.4 Conclusion

Rotation methods have acquired a certain (unwarranted) disreputability because there is a temptation to claim that if the axes in a factor space are successfully interpreted, then there must exist parallel underlying physical processes. It should be emphasized that the axes of a substructure are arbitrary, in that the power of a linear structure to account for a set of observed variables is not related to the location of the axes of that structure. A statistical analysis cannot deduce a 'correct' positioning of the axes; the criterion used for selecting the basis is at the personal choice of the analyst. In practice it is often best to try one or two different rotations in an analysis to give a clearer picture of the relationships involved. The preceding example illustrates the dangers in assigning an underlying physical meaning to arbitrarily located axes. However, used with circumspection, axis rotation clearly provides a deal of useful information, not only in factor and component analysis, but wherever linear combinations of variables can be given as much credence as the variables themselves.

The problems of deciding how much importance to attach to a particular rotation are not lessened by there being no real body of sampling theory for loadings. In only two cases are any results available that make possible significance tests, or even simple measurement of the standard errors of the loadings. The first case is when the rotation is pre-specified; and the second, when the principal axis solution is used (see Section 3.4). In both these cases it is necessary to assume that the observed variables are multi-normally distributed, and to restrict the results to very large samples. Whilst little is known about the adequacy of the results in face of departure from assumption, some work does suggest that the approximations may not be very accurate (see Chapter 6 in Volume 2). Details of the formulae for the sampling errors of the loadings are given in Lawley and Maxwell (1971).

These difficulties notwithstanding, rotation methods are a useful interpretative tool in both PCA and CFA, and it is to the discussion of these two techniques that we now turn.

4.3 PRINCIPAL COMPONENT ANALYSIS

PCA is one of the oldest techniques of multivariate data analysis, and is arguably one of the most useful ways of exploring a covariance structure. With present-day computer technology, and the associated numerical methods that have grown up, it is certainly one of the most frequently used techniques, if only because it is one of the easiest and quickest to apply. As well as having particular, precise uses it provides a valuable set of summary statistics for any correlation (or, more generally, covariance) matrix. If a set of p variables measured on n individuals form the data matrix, $\mathbf{Y}$, then, as previously argued, one might expect the covariance matrix $\mathbf{S}$ to display a pattern of covariation between the variables consistent with some underlying dominant structure. Certainly if, instead, the matrix seemed only to be a random assortment of trivial correlations, then this too would be a matter of interest. The transformation to principal components will highlight linear relationships between the variables, showing both stable colinearities and patterns of covariation. In purely geometric terms, PCA will show how the observed variation spreads through p-dimensional space and to what extent it confines itself to a lower dimensionality.

The structure sought is the occurrence of major dimensions of covariation in the data. To elicit these we consider linear combinations of the variables that reproduce as much as is possible of the original variance in the data. We can always find p such combinations, that is, the principal components, such that

(i) they have no intercorrelations amongst themselves;
(ii) the first q (less than p) of them produce the best q-dimensional reconstruction of the covariance structure of the p variables.
(iii) all p reproduce the covariance matrix completely.

The notion of 'best' in (ii) may be defined in more than one way. The traditional method is to define the jth principal component as a linear combination of the observed variables $\mathbf{C}_j'\mathbf{Y}$ that has maximum variance. To ensure requirement (i) each combination $\mathbf{C}_j'\mathbf{Y}$ is constrained to be uncorrelated with preceding components $1,2,3 \ldots j-1$. Each is also subject to a constraint on the overall size of the coefficients, namely that $\mathbf{C}_j'\mathbf{C}_j = 1$. The constraint on the coefficients is necessary to prevent the variance being arbitrarily increased by making these coefficients ever larger; this restriction ensures that the total variance of all p components is equal to the total variance in the data, in accordance with (iii).

Thus a set of derived variables $\mathbf{C}_j'\mathbf{Y}$ are produced from the data, each in turn having as large a variance as possible subject to (i) and (iii). The first q of these will be the variables which reproduce the maximum variation possible from the original data $\mathbf{Y}$; i.e. they give the best q-dimensional representation of data. This approach is taken up again in Section 4.3.6 from a different standpoint.

An equivalent definition of 'best reconstruction', but one which is often more flexible in its interpretation, can be given as follows: a matrix $\mathbf{S}^*$ is the the best q-dimensional approximation to the covariance matrix $\mathbf{S}$ if it minimizes the sum of squared discrepancies between $\mathbf{S}^*$ and $\mathbf{S}$. (This is usually denoted $\| \mathbf{S} - \mathbf{S}^* \|^2$.) This definition emphasizes two properties of the q-dimensional reconstruction: firstly that it is the covariation $\mathbf{S}$ of the entire set of variables that is being summarized, a point crucial to the analysis; and secondly that the optimality is a property of the q-dimensional space spanned by the components, rather then a property of individual components. This second approach, dealing directly with the covariance in the data, may be extended to deal with situations in which $\mathbf{S}$ is not a proper covariance matrix. It is, in fact, that used in developing *principal coordinate analysis*, a technique due to Gower (1966) which operates on a matrix of distances, rather than on the covariances used in PCA.

Returning to the purely geometric description, consider the variation in the space covered by the observed variables. A PCA program will give the eigenvalues λ_j of the covariance matrix. The largest of these, λ_1, shows the maximum spread of the data in any single direction, or dimension; the next largest, λ_2, shows the spread in the second most important dimension; and so on. The degree to which the data may be accounted for by a few, say q, dimensions may therefore be measured by summing these quantities, $\lambda_1 + \lambda_2 + \ldots + \lambda_q$. This, divided by the total amount of variation in the data, $\lambda_1 + \ldots + \lambda_p$, gives the 'proportion of variance accounted for'. (It is easy to show that $\lambda_1 + \ldots + \lambda_p$ is equal to the sum of the variances of the variables, Σs_i^2). Lack of fit into q-dimensions may thus be measured by the residual eigenvalues $\lambda_{q+1} + \ldots + \lambda_p$. Alternatively the criterion of fit to the covariance matrix given above, $\| \mathbf{S} - \mathbf{S}^* \|^2$, may be used directly. (It may be shown that this is the sum of squared eigenvalues $\lambda_{q+1}^2 + \ldots + \lambda_{p+1}^2$). This is more usefully standardized to a 'proportionate fit' coefficent by using $1 - \| \mathbf{S} - \mathbf{S}^* \|^2 / \| \mathbf{S} \|^2$.

4.3.1 Component loadings

The required vector of weights $\mathbf{C}_j$ for the jth component are obtainable from any standard computer program; these weights are in fact the ortho-normalized eigenvector of the covariance matrix. The weights themselves are used usually only to compute the *component scores*: i.e. the value of the derived component on each of the observations. The vector of scores for the jth component $\mathbf{C}_j'\mathbf{Y}$ may be used in subsequent analysis to represent the position along the jth dimension of the individual observations. The variance of the scores for the jth component is given by the jth eigenvalue λ_j of the covariance matrix. Clearly the size of the variance indicates the component's importance in determining the overall pattern of the covariance structure.

A more useful statistic than the weights $\mathbf{C}_j'$ is the vector of loading $\mathbf{L}_j$ for the jth component across the variables; these are simply the weights rescaled by the component's standard deviation, i.e. $\lambda_j^{1/2}\mathbf{C}_j$. The loadings thus reflect not only the pattern of contribution of the variables to the component, but also the relative importance of each component in the overall covariance structure. It should be noted that in the case where a correlation matrix has been analysed, these loadings are also the correlations of the variables with the components (for this to be true generally each coefficient must be divided by the standard deviation of the relevant variable, s_i). For these reasons it is the loadings matrix $\mathbf{L}$ that plays the important role in the analysis, providing as it does comparability across both variables and components.

The complete set of components is seldom used in subsequent analysis, some decision being taken to deal with only the first few (see Section 4.3.4). The first q component loadings $\mathbf{L}$, as derived directly from the eigenvectors $\mathbf{C}$, constitute a principal axis rotation (see Section 4.2.1) for the q-dimensional component space. However, any rotation procedure may be applied, as discussed in Section 4.2, to give a set of 'rotated principal components' spanning the same space. The proportion of total variance accounted for by the subspace is not changed, but a new positioning of the axes will change the variance of the (now rotated) components, and so their importance relative to each other. Note that a rotation produces completely new components; there is no sense in which the first rotated component, for example, has an affinity with the first raw component, rather than the second or any other. In this respect 'axis rotation' is a misnomer; 'setting new axes' would be a more accurate term. If the rotation is to an oblique set of axes, then property (i) above, that the components are uncorrelated, will no longer hold: the rotated component scores will be correlated with each other to the same extent that the axes are oblique. Under these circumstances the variance of the rotated components do not constitute a partition of the total variance attributable to the component space, and care should be taken in their interpretation. The situation is exactly analogous to regression analysis with correlated regressor variables, where the variance due to regression cannot be partitioned uniquely amongst the regressors.

In using the only the first few components, the extent to which a variable can be fully represented by this subspace is measured by the SMC with the subset. It is convenient to introduce a term from factor analysis and refer to this as its (squared) communality with the component subspace. When **S** is a correlation matrix these values are those in the diagonal of the approximation **S***. Generally, however, the diagonal element must be taken in ratio to the original variance of the variables to give the SMC. In the loadings matrix **L** a variable with high communality will show high (positive or negative) values; if the rotation is an orthogonal one, the squared communality will in fact be the sums of its squared loadings on the components. Thus to have a high communality and be well-represented by a set of (rotated) components a variable must either load highly on one, or moderately highly on several different components.

*The characterization of a component as a combination $\mathbf{C}_j'\mathbf{Y}$ having maximum variance is derived by maximizing

$$\mathbf{C}_j'(n^{-1}\mathbf{YY}')\mathbf{C}_j \quad \text{subject to} \quad \mathbf{C}_j'\mathbf{C}_j = 1.$$

(No loss of generality occurs from standardizing the data matrix **Y** to have zero mean.) Using Lagrangian Multiplier $-\lambda_j$ we obtain, after differentiating,

$$(n^{-1}\mathbf{YY}')\mathbf{C}_j = \lambda_j\mathbf{C}_j \tag{1}$$

This defines $\mathbf{C}_j$ as an eigenvector of the covariance matrix; and their orthogonality follows from the matrix's symmetry. The variance of the component is

$$n^{-1}(\mathbf{C}_j'\mathbf{Y})(\mathbf{C}_j'\mathbf{Y}) = \lambda_j$$

from premultiplying (1) by $\mathbf{C}_j'$.

For the definition of PCA in the text we use the Singular Value Decomposition. If a matrix **M** has the decomposition

$$\mathbf{M} = \sum_j \mathbf{C}_j\mathbf{B}_j'v_j, \tag{2}$$

that is, for $j = 1, 2, 3, \ldots$

$\mathbf{C}_j$ are the ortho-normal eigenvectors of $\mathbf{MM}'$

$\mathbf{B}_j$ are the ortho-normal eigenvectors of $\mathbf{M}'\mathbf{M}$

v_j^2 are the (ordered) eigenvalues of $\mathbf{MM}'$ or $\mathbf{M}'\mathbf{M}$.

then the principal theorem states that the least squares q-dimensional approximating matrix **M*** is formed by the first of terms of the summation in (2). The proof follows from the Courant–Fisher characterization of an eigenvalue of an $n \times n$ matrix $\mathbf{A}$ as

$$\lambda_j = \min_{(\text{all } \mathbf{Z} \text{ rank } j-1)} \max_{(\text{all } \mathbf{x} \perp \mathbf{Z})} \{\mathbf{x}'\mathbf{A}\mathbf{x}\}$$

where $\mathbf{x}$ is $n \times 1$, $\mathbf{Z}$ is $n \times (j-1)$.

We note that substituting $\mathbf{S}$ for $\mathbf{M}$ in (2) we have immediately

$$\mathbf{S}^* = \sum_j^q \lambda_j \mathbf{C}_j \mathbf{C}_j' \qquad (3)$$

where $\mathbf{C}_j$ are the eigenvectors of $\mathbf{S}$, $\mathbf{S}\mathbf{S}'$ and $\mathbf{S}'\mathbf{S}$; λ_j are the eigenvalues of $\mathbf{S}$.

Finally, noting that λ_j, $\mathbf{C}_j$ for $j = 1, 2, \ldots q$ are the eigenvalues and eigenvectors of $\mathbf{S}^*$, it follows immediately from the trace of a matrix being the sum of its eigenvalues that

$$\begin{aligned} \| \mathbf{S} - \mathbf{S}^* \|^2 &= \mathrm{tr}(\mathbf{S} - \mathbf{S}^*)(\mathbf{S} - \mathbf{S}^*) \\ &= \mathrm{tr}(\mathbf{S}^2) + \mathrm{tr}(\mathbf{S}^{*2}) - 2\mathrm{tr}(\mathbf{S}\mathbf{S}^*) \\ &= \lambda_{q+1}^2 + \ldots + \lambda_p^2 \end{aligned}$$

4.3.2 Example 2

The robustness of the PCA approach, that is, the least squares approximation to a covariance matrix, makes it a useful technique in non-standard situations, where the data may not fit classical assumptions of multinormality. Here we deal with a set of data collected in a study of the symptomatology of alcoholism (Hershon, 1975). The incidence of various symptoms in 100 diagnosed alcoholics was recorded and, after extensive transformation, a data matrix produced which consisted of 36 symptoms. From this a covariance matrix was computed for the symptoms and subjected to a PCA. One would not usually take principal components of a covariance matrix that had less than 3 or 4 times as many observations as variables. In this case, however, it is not unjustifiable, since the 36 rankings were based on 100 initial observations and might be expected to be reasonably stable.

Table 4.2 shows the matrix fit coefficient $1 - \| \mathbf{S} - \mathbf{S}^* \|^2 / \| \mathbf{S} \|^2$ in column 4 and the proportion of variance accounted for in column 3. It can be seen that the former is simply an exaggeration of the first. On either reading the first component is strong, suggesting an overall high correlation in symptom incidence: they all occur or none occur. Through this over-riding pattern there is some indication of a second component but no other component is prominent. On this basis the loadings of the first two varimax rotated components were computed, and the squared communality of each variable with the component space. From these latter, five symptoms have very little in common with the major pattern of variation (3, 12, 13, 18, 31). The remainder, with four exceptions, fall neatly into groups as indicated by their loadings: the majority are characterized by a high loading only on the first axis, and most of the remainder load only on the second; two (30, 35)

Table 4.2. Component analysis of 36 symptoms (Example 2). Varimax rotation of the first two principal components. (Loadings within ±0.25 have not been shown)

Component number	Variance	Proportion of total	Cumulated proportion	Fit to matrix	Symptom number	1st loading	2nd loading	Communality squared
1	13.25	0.37	0.37	0.73	1	0.82		0.78
2	6.04	0.17	0.54	0.88	2	0.75		0.60
3	3.38	0.09	0.63	0.93	3			0.11
4	2.50	0.07	0.70	0.96	4		0.62	0.50
5	2.22	0.06	0.76	0.96	5	0.74		0.59
6	1.51	0.04	0.80	0.97	6	0.64		0.41
7	1.37	0.04	0.84	0.97	7	0.81		0.72
8	0.93	0.03	0.87	0.98	8	0.84		0.71
9	0.89	0.02	0.87	0.98	9	0.68		0.51
10	0.54	0.02	0.91	0.99	10		0.83	0.70
11	0.50	0.01	0.92	0.99	11	0.81		0.70
12	0.43	0.01	0.93	0.99	12			0.02
13	0.35	0.01	0.94	0.99	13			0.22
14	0.28	0.01	0.95	1.00	14		0.63	0.42
15	0.26	0.01	0.96	—	15	0.62		0.43
16	0.22	0.01	0.96	—	16		−0.46	0.36
17	0.21	0.01	0.97	—	17	0.92		0.87
18	0.20	0.01	0.97	—	18			0.03
19	0.16	0.00	0.98	—	19	0.72		0.54
20	0.13	—	0.98	—	20		0.81	0.69
21	0.12	—	0.99	—	21	0.77		0.62
22	0.10	—	0.99	—	22		0.53	0.30
23	0.09	—	0.99	—	23	0.77		0.60
24	0.08	—	0.99	—	24	0.77		0.63
25	0.06	—	0.99	—	25		0.82	0.71
26	0.05	—	1.00	—	26	0.75		0.69
27	0.04	—	—	—	27		0.72	0.58
28	0.03	—	—	—	28	0.70		0.62
29	0.03	—	—	—	29	0.91		0.84
30	0.02	—	—	—	30	0.66	0.58	0.77
31	0.02	—	—	—	31			0.02
32	0.01	—	—	—	32		−0.50	0.34
33	0.01	—	—	—	33	0.78		0.65
34	0.00	—	—	—	34	0.83		0.73
35	0.00	—	—	—	35	0.51	0.54	0.55
					36		0.76	0.77

load on both, and two (16, 32) have negative loadings on the second axis. The communalities of the variables in the two groups are satisfactorily high, with the exception of 6, 14, 15 and 22, and it can be argued that a two-dimensional solution will represent the other symptoms. It happened that the symptoms loading on the first axis were all symptoms of a psychological nature, and those loading on the second were all physical disturbances of some kind. These initial observations prompted a cluster analysis of the symptoms, which subsequently confirmed the classification suggested by this analysis.

4.3.3 Variance Profiles

A useful alternative in interpreting PCA is provided by viewing it as a technique for constructing what might be termed variance profiles. The scores in each column of the data matrix **Y** give a profile of that observation across the variables. We seek a single profile $\mathbf{P}_1$ that best represents all observation profiles, in the following sense:

(i) that each observation reproduces the rise and fall of the profile uniformly, i.e. to within a scaling factor (see Figure 4.3)

(ii) that the total squared deviation from the (scaled) profile, when summed across all observations, is minimal.

Since one profile is not usually a good representation of all the observed profiles, the analysis may be extended to allow the deviations from this profile to be fitted by a subsequent profile; and so on for as many profiles as seem desirable, $\mathbf{P}j$ for

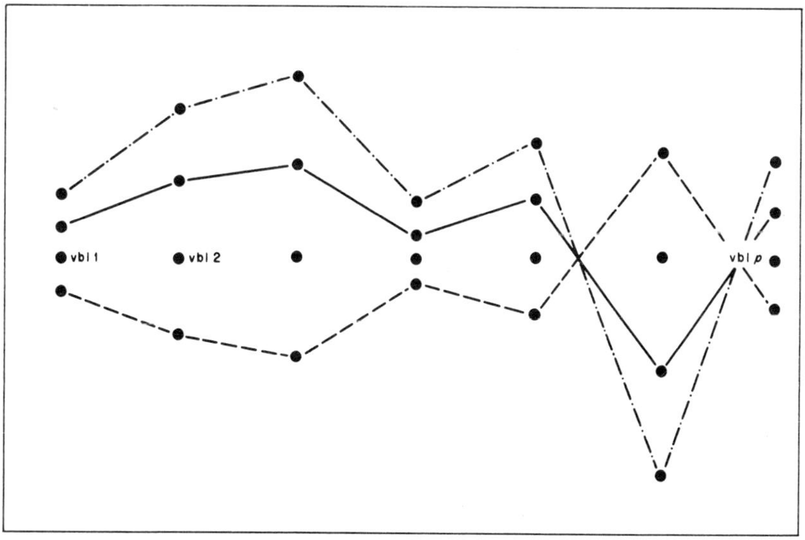

Figure 4.3. An example of a profile across 7 variables showing (i) basic profile (——), (ii) with negative scaling factor (– – –) and (iii) with larger (positive) scaling factor (. _ .)

$j = 1, 2, 3, \ldots$. Note that the scaling factors used to fit each observation profile are picked automatically to give the best fit.

Two points need emphasis here. First, the profile can occur in varying degrees, from observation to observation, but the pattern of the profile remains stable. The scaling factor, or score, with which the ith observation loads on the jth fitted profile (denoted f_{ji}) is not confined to positive values. By allowing a negative value the profile is 'reflected' for that observation (see Figure 4.3). For this reason it is probably easier to interpret the results if the data matrix is first standardized to give each variable a zero mean; under these circumstances the profiles portray the pattern of variation about the mean.

The second point to be emphasized is that in the case of several fitted profiles, we do not confine one group of observations to one profile, and a second group to another; each observed profile *may* be a composite of several fitted profiles. In this respect a distinct difference occurs between this analysis and cluster analysis. In the latter, in its usual form, observations are matched to groups each having a single profile. In interpreting the profiles it must be remembered that the order of variables within the profile is of course arbitrary.

The connection with PCA may now be made explicit. The profiles $\mathbf{P}j$ are identical with the principal component loadings $\mathbf{L}_j$, and the scaling factor f_{ji} for the jth profile on the ith observation is the corresponding principal component score for the observation. By reconstructing each the observation from, say, q fitted (scaled) profiles an approximation

$$\mathbf{Y}_i^* = \mathbf{P}_1 f_{1i} + \mathbf{P}_2 f_{2i} + \ldots + \mathbf{P}_q f_{qi} = \mathbf{P}\mathbf{f}_i$$

to the data matrix $\mathbf{Y}$ is formed. Since summing the discrepancies of each observation's profile $\| \mathbf{Y}_i - \mathbf{Y}_i^* \|^2$ for all observations, as dictated in (ii), is the same as summing over all elements of the data matrix, written $\| \mathbf{Y} - \mathbf{Y}^* \|^2$, it can be seen that $\mathbf{P}$ and $\mathbf{F}$ are chosen to give a least squares fit to the data $\mathbf{Y}$. Thus PCA produces the best q-dimensional reconstruction of both the covariance matrix and the data matrix, a feature which makes it a fairly robust analysis. Note that the profiles may be subjected to arbitrary rotation as may component axes, without changing their optimality properties.

*The definition of a set of profiles $\mathbf{P}_j$, $j = 1, 2, \ldots q$, requires the total sums of squares of the residuals over all observations to be minimal. For the ith observation $\mathbf{Y}_i$ the residuals are

$$\mathbf{Y}_i - \mathbf{P}_1 f_{1i} - \mathbf{P}_2 f_{2i} - \ldots \mathbf{P}_q f_{qi}$$

and denoting

$$\hat{\mathbf{Y}}_i - \sum_j^l \mathbf{P}_j f_{ji} \tag{4}$$

the sum of squared residuals is $(\mathbf{Y}_i - \hat{\mathbf{Y}}_i)'(\mathbf{Y}_i - \hat{\mathbf{Y}}_i)$. Thus the total over all observations is

$$\sum_i (\mathbf{Y}_i - \hat{\mathbf{Y}}_i)'(\mathbf{Y}_i - \hat{\mathbf{Y}}_i) = \text{tr}(\mathbf{Y} - \hat{\mathbf{Y}})'(\mathbf{Y} - \hat{\mathbf{Y}})$$

It is obvious that $\mathbf{Y}^*$ is of rank q, and so by applying the approximation theorem, as in (2) and (3), must be the sum of the first of terms of the Singular Value Decomposition of $\mathbf{Y}$. We may write (4) as

$$\mathbf{Y}^* = \sum_{j}^{q} \mathbf{P}_i \mathbf{F}_j^T$$

where $\mathbf{F}_j^T$ is used to denote the jth column of the transposed matrix $\mathbf{F}^T$ and this must hold for $q = 1, 2, \ldots, p$; it follows that $\mathbf{P}_j\mathbf{F}_j^T$ is the jth term in that decomposition. The scaling indeterminacy is resolved by arbitrarily choosing

$$\mathbf{P}_j = v_j \mathbf{C}_i, \quad \mathbf{F}_j^T = \mathbf{B}_j, \quad v_j = \lambda_j^{1/2}$$

in the decomposition of $\mathbf{M} = \mathbf{Y}$.

That $\mathbf{F}_j^T$ are the standardized component scores is easily shown by

$$\mathbf{C}_j'\mathbf{Y} = \mathbf{C}_j'(\sum_j v_j \mathbf{C}_j \mathbf{B}_j') = \mathbf{B}_j' = \mathbf{F}_j^{T\prime}.$$

4.3.4 Example 3

This example uses the same data as Example 1 (see Fig. 4.1 and Table 4.1), is taken from a study of attitudes towards cigarette smoking by Russell, Peto and Patel (1974). A sample of 420 people gave information on a number of attitudes, of which a subset is given in Figure 4.4. The data, measured on a 4-point scale, did not seem fine enough to support a detailed analysis of its error structure, as would be required by CFA (q.v. Section 4.4). In this study a PCA might have been carried

Vbl	Profile I	Profile II
13	0.33	0.43
14	0.26	0.37
15	0.02	0.52
16	−0.10	0.30
17	0.05	0.45
18	0.10	0.27
19	−0.17	0.10
20	0.13	0.18
34	0.26	0.30
30	0.49	0.28
31	0.49	−0.13
1	0.61	−0.18
2	0.53	−0.02
3	0.72	0.07
4	0.51	0.14
5	0.50	0.09
6	0.59	−0.21

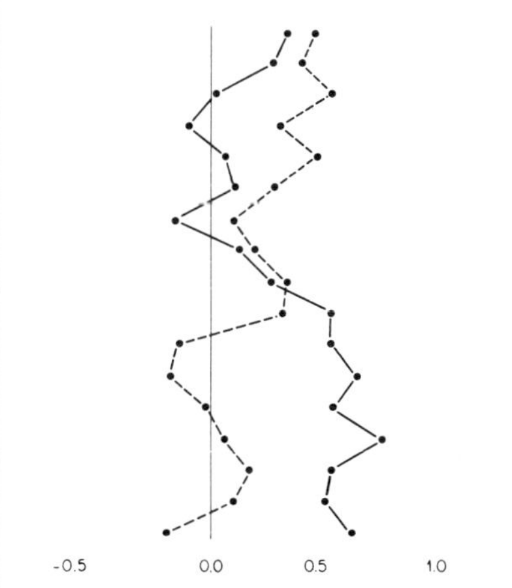

Figure 4.4. Study of a subset of data for Example 3.
Key — continuous line: Profile I, broken line: Profile II.

out on the covariance matrix, since all variables were measured on the same scale, but it was nevertheless decided to standardize all variables to zero mean and unit variance. It can be argued that this has the effect of removing the observed differentials in vehemence and indifference from the responses.

The eigenvalues from the correlation matrix show a pattern typical of this type of data: a fairly large first eigenvalue which accounts for 21 per cent of the variation; a second and third accounting for 11 per cent and 8 per cent and the remainder tailing away from 5 per cent, with a large number (three quarters of them) each accounting for less than 3 per cent. A 1-, 2- or 3-dimensional approximating space might have been fitted, and the results of a 2-dimensional oblimin fit are shown in Figure 4.4. We note that this is no more than a graphical representation of the rotated component loadings, and largely supports what (in this case) can be seen in Table 4.1 above.

The authors point out that in this study the PCA gave approximately the same solution as did CFA. Subsequently oblimin scores were computed for the two profiles, and those for the first profile were found to correlated to a high degree with cigarette consumption. Those for the second were also slightly correlated, and, as expected in an oblique rotation, correlated with the first profile, $\rho = 0.14$.

4.3.5 Number of Components

A question deliberately avoided in the foregoing sections is that of how to determine the number of components that are needed. The examples have implied that we wish to account for the major part of the total variation in the data (say 60 per cent or 70 per cent); or that we wish to use only those major components, whose variance stands out as far larger than the rest (say 30 per cent greater); or that a number considered reasonable *a priori* be chosen. The first two of these requirements are simply rules of thumb that the analyst might choose. They cannot be usefully formulated as hypotheses to be tested by the data, and so represent a method that is open to considerable arbitrariness. However; in practice these are often the most useful approaches to the problem. The third approach, of choosing the number of components in advance of the analysis, can also be seen as a purely arbitrary decision. However, it is possible to treat it as an hypothesis to be assessed in light of the data, although here difficulties arise in formulating the exact structure being hypothesized.

Kendall and Stuart (1961) have pointed out that an hypothesized low-dimensional multivariate linear structure cannot possibly give rise to an observed structure of higher dimensionality. Thus a direct test on the rank of the covariance matrix is pointless. It is necessary to superimpose on the low-dimensional structure some hypothesized error structure that leads to the observed data. The usual formulation postulates that no correlations at all can be found amongst (orthogonal combinations of) the residuals from the larger components. This is equivalent to postulating that the remaining components have equal variances (within sampling error), i.e. that no one stands out above the others. This is the hypothesis of residual isotropy or sphericity, which can be useful in some circumstances (but see section 4.3.6). The test statistic used is the likelihood ratio statistic, and its sampling distribution is known only for large sample sizes from a normal distribution. The way it behaves in small samples or in non-normal data is not certain. Some work (see chapter 6 of Volume 2) sheds considerable doubt on its validity, but other authors, e.g. Lawley and Maxwell (1971) recommend its use as long as the sample size is fairly large, for instance $n - p \geqslant 50$. The statistic is computed from the eigenvalues as

$$n[(p-q)\log \sum_{q+1}^{p} \lambda_j (p-q)^{-1} - \sum_{q+1}^{p} \log \lambda_i]$$

and is usually referred to a χ^2 – distribution with ½ $(p - q + 2)(p - q - 1)$ degrees of freedom. In its extreme form with q equal to zero, the sphericity hypothesis is equivalent to the hypothesis that all correlations are zero, within sampling error, and the test statistic becomes

$$-n \log |S|$$

In almost any reasonable data set, however, this latter hypothesis will generally be found unacceptable.

4.3.6 Applications

PCA is mainly used as a tool for data reduction: that is, as a transformation which shows which parts of the data can be discarded with little loss of information. By showing which linear functions of the variables can be ignored (namely, as discussed above, the smaller components) it can be used to reduce and data set to a few major, summary items. In addition, it can be used to show which variables can be omitted from the set without changing too much the information base. Jolliffe (1973), for example, discusses its use in this fashion and compares it with multiple regression techniques used for the same purpose. An example using all p principal components is provided by Andrews' (1969) technique for diagrammatic representation of a data matrix. He recommends rotating to principal components so that the salient points in the diagram he uses can be matched to the salient sources of variation in the data. The use of PCA in regression techniques is common. Daling and Tamura (1970) use varimax rotated components of the regressor variables in stepwise regression to limit collinearities amongst the variables which are chosen to be included in an efficient prediction equation. Hawkins (1973) uses a PCA of both regressors and regressand in a similar fashion. By inspecting the collinearities, alternative equally acceptable regressor sets can be identified. Hawkins (1974) also finds a use for the smallest components as detectors for outliers or errors in general data sets.

In using PCA as an exploratory tool in its own right it is tempting to identify major axes of variation with underlying 'physical' processes. Kendall and Stuart (1967) report and comment upon a study by Craddock involving a time series of monthly temperatures in England. The example is interesting because it not only succeeds in giving a substantive interpretation of the principal axis solution but also because, in part, it uses a data matrix *not* standardized to zero mean. In other studies, in questionnaire analysis for example, PCA is used as an alternative to CFA. The first few rotated components often give a robust identification of major trends in the data, which may be treated as scales and indices of underlying properties. The profile interpretation of the component loadings has been discussed in detail here because it affords an intuitive insight into both advantages and disadvantages of this type of analysis. Its main strength is in its robust, least squares approximation to the data matrix (and additionally to the covariance matrix). It is easy to see the direct effect on the profile $\mathbf{P}_j$ of using less than ideal measurements, for instance binary data or ipsative scores or heavily skewed data. It also points up the difference between data suitable for cluster analysis (of observations) as opposed to component analysis: in the latter an observation may be a composite of profiles; in the former several observations must share a single common profile.

There are three serious disadvantages with PCA. Most immediately, PCA is dependent upon the scale of the measurements, and is best used when some common natural scale is intrinsic to all the variables being analysed. If the total squared residuals $\| \mathbf{Y} - \mathbf{Y}^* \|^2$ is divided into subtotals for the fit of each variable separately,

$$nv_j^2 = \sum_i (y_{ji} - \hat{y}_{ji})^2$$

then it is clear than an overall minimum will be attained by, *ceteris paribus*, fitting closely the variables with comparatively large variances; these variables will thus dominate the analysis.

The common practice of using the correlation matrix and not the covariance matrix ensures that each variable carries equal weight in this respect, by relating its unaccounted variation to its total variation; that is, $\Sigma_j(v_j^2/s_j^2)$ is minimized. Some further justification can be found by noting that this is equivalent to maximizing the total squared communalities, or SMC's, of the variables (see Section 4.2.1) and that these are scale-free quantities.

The second point is that PCA taps the joint variation in a set of variables. If several similar variables are included in the set, they will dominate the components in a fashion similar to one variable with a large variance; as the composition of the set shifts emphasis from one type of variable to another, so will the completeness with which they are represented by the components. This problem occurs again in all factor analysis techniques (see Section 4.4.4).

Thirdly, the hypothesis of residual isotropy (see Section 4.3.5) is not altogether satisfactory as a formulation of the concept of no residual structure. The eigenvalues λ_j, as well as the eigenvectors, are sensitive to changes in scale of the variables; and if the lower eigenvalues $\lambda_{q+1}, \ldots, \lambda_p$ are equal under one scaling, they will not be so under another.

It is through the attempt to deal with the difficulties involved in scaling the variables and in formulating a satisfactory error structure that CFA models arise and it is from this angle that the next section approaches the subject.

4.4 COMMON FACTOR ANALYSIS

In Section 4.3.5 it was noted that PCA leads to difficulties in specifying when 'all' the structure in a set of variables has been identified, and only random noise remains. The CFA model solves this problem by requiring that this noise consist of random residual variables having particular testable properties, namely, that they are not only uncorrelated with the identified structure, but in addition have no intercorrelations amongst themselves. Thus, being totally unrelated to everything else, they might reasonably be termed simply 'noise'. When an approximating space is found that leaves residuals with these properties, all the common structure in the variables has been found, and the approximating space is termed the common factor space. Howe (1955) formulates the problem in terms of partial correlations, to the same effect: can a set of q unobserved factors be fitted, such that the correlations between the variables, after the effect of these factors has been 'partialled out', are all zero. It is thus usual to see what *minimum* number of factors, q, is required to achieve this effect.

If the variables all seek to measure the same underlying phenomenon, or are all in part caused by it, then the model is an intuitively reasonable formulation. The residuals may be treated as errors of measurement acquired in trying to observe a 'true' underlying variable, as for example in some intelligence testing methods. Another useful interpretation is to regard each variable as having an element particular to itself and separate from the central phenomenon; then the residuals may be treated as 'unique' effects of the measured variables, in contrast to the 'common' effects of the factors. The latter conceptualization has led to the proportions of a variable's variation attributable to each part being called its *uniqueness* and *communality*, respectively.

Thus in a common factor analysis one needs to fit, say, q (less than p) factors to a set of p variables, and tries to find:

(i) a q dimensional covariance structure (derived from the common factors); and
(ii) a set of p error variances (i.e. a diagonal error covariance matrix)

that together reproduce the observed covariance matrix, S. In separating out the q-dimensional common factor structure from the p unique variances, it should be noted that this is always possible as a fact of simple algebra, whatever the observed covariances S, if $(p-q)^2 = p + q$; and if $(p-q)^2$ is less than this there are an infinte number of possible solutions. The only non-trivial circumstance is when the number of factors is considerably less than the number of variables, i.e. $(p-q)^2 > p + q$. Some of the critical values are given in Table 4.3, and it is usually recommended that in practice the number of variables be at least 5 times the number of factors.

Table 4.3. Number of factors allowed for a non-trivial factor structure

No. of variables, p	3	4	5	6	7	8	9	10	15	20
No. of factors, q	0	1	2	2	3	4	5	5	9	14

In the non-trivial case, it is not generally possible to satisfy exactly both the above requirements simultaneously, i.e. that (i) the common factor structure is q-dimensional, and (ii) that the error covariance matrix is zero apart from the diagonal elements. In any given set of data the closest solution is needed, and the two most important approaches to this optimization problem give the same solution. Rao (1955) considered fixing the error matrix at diagonality, but choosing the values in the diagonal to make the remaining common structure as near to q-dimensional as possible; his criterion was to maximize the q canonical correlations between the observed variables and the factor space. Lawley and Maxwell (1963, 1971) constrained the factor space to be q-dimensional and made the error covariance matrix as near-diagonal as possible; to do this they assumed the error variables distributed in the multivariate normal form, and maximized the likelihood criterion. Since both formulations provide the same solution, Rao's canonical factor approach shows that the solution does not need the assumption of normality in the variables, but requires the data analyst only to justify using product-moment correlations. However, the advantages of assuming normality are two-fold. Firstly, that standard errors can be computed for the resulting maximum likelihood estimates under this conditon. Secondly, an overall test of significance may be made (using the likelihood ratio criterion) for the hypothesis of no residual intercorrelations. These two points are taken up in Sections 4.4.1 anf 4.4.2.

Whichever approach is taken, the result is a decomposition of the observed covariance matrix, S, into two matrixes representing the common and unique elements of covariation.

$$\mathbf{S} = \mathbf{S}^* + \mathbf{V}$$

The designating of the common covariance contribution as $\mathbf{S}^*$ is made mainly to keep the parallel with PCA notation in Sections 4.1 and 4.2. It is loosely justifiable in that it represents a reconstruction of the covariances between the variables, were there no errors.

There are three features of CFA which bear comment at this juncture. Firstly, the form of the solution itself, algebraically and numerically, is very similar to a q-dimensional PCA solution of a particular rescaling of the data. To be precise, if the data are scaled to have equal error variances, i.e. the jth variable is weighted by v_j^{-1} to give a covariance matrix $\mathbf{V}^{-1/2}\,\mathbf{S}\mathbf{V}^{-1/2}$, then each factor is simply a rescaling of the appropriate principal component. Thus all the considerations that apply to interpreting the statistics from a PCA apply to CFA, too. In particular communalities, uniqueness, loadings all have the same interpretation in both analyses. However, it should be pointed out that the computation of the error variances needed for the rescaling of the covariance matrix in CFA is extremely difficult; not until papers by Joreskog (1967) and Clarke (1970) were efficient and successful computational procedures available. It is this factor, that the error covariances are more difficult to compute than is at first apparent, which in the past has produced a number of numerical procedures purporting to give CFA solutions, but which in practice are often not useful.

Secondly, the rescaling procedure described above is such that the solution is independent of the original scale of the variables. This arises because the formulation of the structural properties required of the solution is solely in terms of correlations (see above), and these are scale-free statistics. By contrast, as pointed out in Section 4.3, PCA in its general formulation looks to reconstruct observed variances, which obviously depend heavily on the scale in which the variables are measured. This scale-invariance is possibly the most useful single feature of CFA in practice. For it is seldom the case that a set of variables are meaningfully measured on jointly compatible scales and without this, *a priori* decisions about their relative scaling would have to be made.

Thirdly, it is to be emphasized that whilst the forms of the PCA and CFA solutions are similar, they are based on totally different principles. In PCA one minimizes only the size of the residual errors, and specifies nothing about their intercorrelations; in CFA it is only their intercorrelations that are minimized. Both analyses produce residuals that are uncorrelated with the approximating space. It should be noted that the characterization sometimes attributed in casual discussion to CFA: that it tries to explain as much of the data as possible by breaking down the variables into orthogonal factors, is inaccurate. As is pointed out in Section 4.2.1, oblique axes for the factor space are a natural as orthogonal axes from the stand-point of the statistical analyses. It is uncorrelated *errors* that CFA fits and it is their intercorrelations that are minimized, rather than their overall size. However, the similarities in form between PCA and CFA do often lead to similar results in a given data set. One can expect this to occur when there is a fairly clear CFA structure – that is, the error variables actually occuring in the observations are reasonably uncorrelated, so that a good fit is obtained – and particularly when the communalities of the variables are approximately equal.

4.4.1 Number of factors

As with PCA there are two aspects to determining how many factors to fit to the data, apart from having *a priori* knowledge of the appropriate number of factors. One may decide from a simple heuristic inspection of the fit obtained; alternatively one may assume multi-normality for the distribution of the data and use a statistical test of hypothesis.

Before the maximum likelihood CFA algorithms became widely available, many factor analysis numerical methods were used which did not necessarily supply good approximations to the likelihood solution. This led to the development of many 'rules of thumb' for determining the number of factors required. The considerations in CFA for determining this number have slightly different emphasis from those used in PCA. The formulation of CFA demands that it is the absolute size of the factors (i.e. the variance) that is important whereas in PCA it is the proportion of the total variation accounted for which is the key measure. Foremost amongst these rules of thumb is Guttman's (1954) strongest algebraic lower bound: the minimum number of factors required is equal to the number of eigenvalues greater than unity in the correlation matrix. The usefulness of thes rule is severely limited because it ignores all sampling fluctuations that arise in the data. Since it is an algebraic solution to the problem, the slightest suggestion of the presence of a factor requires including that factor in the analysis, even if it could be explained by small sampling deviations from a simpler model. (Guttman originally presented this rule as applying to a correlation matrix having no sampling errors.) Harris (1962) and Kaiser (1963) in following Guttman's (1953) Image Analysis approach have argued that one should not use *more* factors than the Guttman rule suggests. Kaiser (1970) reports moreover that the rule itself supplies a number which often is near a half the number of variables used, and that this is too many for practical analysis. He further conjectures that a better rule is to use not unity, which is the average of the diagonal elements of the correlation matrix, but the average of the diagonals in the rescaled correlation matrix $\mathbf{V}^{-1/2}\mathbf{S}\mathbf{V}^{-1/2}$. To be accurate, Kaiser suggests using not the quantities v_j, which would require computation at each value of q, but an approximation $u_j = s_j\ (1 - R_j^2)$ where R_j^2 is the SMC of the j variable on the remaining $(p - 1)$ variables. The u_j so defined are the upper bounds on the uniquenesses v_j given by Guttman (1954). Thus a factor would be fitted if the corresponding eigenvalue were greater than $p^{-1}\Sigma_j(s_j^2/u_j^2)$.

By assuming multinormality an approximate significance test becomes available for large samples. Based on the likelihood ratio test, we have the criterion

$$n[\log |\mathbf{S}^* + \mathbf{V}| - \log |\mathbf{S}|]$$

which is distributed as χ^2 with $\frac{1}{2}[(p - q)^2 - (p + q)]$ degrees of freedom. Box (1949) showed that a better approximation is achieved by adjusting n in the formula to $n\ (2q + 5)/6$; and Bartlett suggested that further accuracy might be obtained by using instead $n - (2p + 4q + 5)/6$. Lawley and Maxwell suggest that the approximate test be used when $n - p \geqslant 50$, assuming multinormality. Thus the procedure usually recommended is: fit a small number of factors; test for the adequacy of the fit with the likelihood-ratio test; if rejected, increase the number of hypothesized factors by one; and repeat. Note that it is necessary to repeat the fitting of the model each time a new number of factors is hypothesized. In this respect there is an important practical difference from PCA, where the number of components wanted can be gauged from the eigenvalues of a single analysis.

A further issue complicates the determining of the number of factors used in a CFA, namely the existence of what are termed *improper solutions.* It often happens in an analysis that fitting a particular number of factors to the covariance structure results in some of the variables being estimated to have zero uniqueness. This implies that they may be entirely accounted for by the common factor structure and contain no error variation of any sort. Whilst this raises no new conceptual problems, the frequency with which the result occurs in practice has given rise to some concern. It can be argued that the occurrence of variables which have no measurement error associated with them is *a priori* unreasonable, especially if information about the set of variables is available from other studies. Deleting the offending variables from the analysis is an unsatisfactory solution, and indeed sometimes results only in the impropriety being transferred to other variables. Reducing the number of factors being used to account for the data is an obvious response, but is not appropriate if the significance testing procedure rejects the hypothesis of fewer factors. This in turn has led to concern that the significance test might be hypersensitive to small non-linearities in the data and so demand that a large number of factors be fitted. Taken in conjunction, these two points cast some doubt upon the robustness of the maximum likelihood estimation procedure. Little is known theoretically on this matter, but the empirical study by Joreskog (1967) is of interest, where eight out of eleven sets of data yielded improper solutions.

The occurrence of zero uniquenesses also complicates the computational procedures involved in deriving the CFA solution. The technique that Lawley and Maxwell (1971) and Joreskog (1967) recommend is to set aside the variables so identified, to run a subanalysis on the remainder using correlations partialled on those variables removed, and then incorporate those variables with the derived factors into a final solution. Fortunately this procedure is automatic in the published computer algorithms. It is described in detail in, for instance, Lawley and Maxwell (1971).

4.4.2 Example 4

The example used is a troublesome set of data which Maxwell (1961) analysed. He encountered some difficulty caused by the comparatively unsophisticated nature of the numerical algorithms then available, and the data were subsequently analysed by Joreskog (1967) using the technique for handling improper solutions described in Section 4.1. Details of each stage of the numerical computations are given in both the Joreskog paper and Lawler and Maxwell (1971). It should not be presumed that CFA always produces such complications.

Table 4.4 gives the correlations between 10 ability scores, derived from a survey of 810 children. The first five of these scores are (1) verbal ability, (2) spatial ability, (3) reasoning, (4) ability, and (5) verbal fluency. The remaining scores Maxwell describes as inventories assessing orectic tendencies: (6) neuroticism, (7) ways to be different, (8) worries and anxieties, (9) interests and (10) annoyances. Note that since CFA is scale-invariant it does not matter whether

Table 4.4. Covariance matrix for Example 4 (Reproduced with permission from A. E. Maxwell, *Journal of the Royal Statistical Society, Ser. A*, **124**, 49–59 (1961))

1.	2.	3.	4.	5.	6.	7.	8.	9.	10.
1.000	0.345	0.594	0.404	0.579	0.280	0.449	0.188	0.303	0.200
	1.000	0.477	0.338	0.230	0.159	0.205	0.120	0.168	0.145
		1.000	0.498	0.505	0.251	0.377	0.186	0.273	0.154
			1.000	0.387	0.168	0.249	0.173	0.195	0.055
				1.000	0.151	0.285	0.129	0.159	0.079
					1.000	0.363	0.359	0.227	0.260
						1.000	0.448	0.439	0.511
							1.000	0.429	0.316
								1.000	0.301
									1.000

Table 4.5. Maximum likelihood loadings for Example 4. (Unrotated combined loadings) (Reproduced with permission from A. E. Maxwell, *Journal of the Royal Statistical Society, Ser. A*, **124**, 49–59 (1961))

Vbl	Factor 1.	2.	3.	4.	Uniqueness
1	0.75	0.03	−0.11	0.19	0.39
2	0.47	0.10	0.36	0.12	0.62
3	0.77	0.17	0.22	0.19	0.30
4	0.53	0.20	0.12	0.17	0.64
5	0.67	0.25	−0.35	0.13	0.35
6	0.26	−0.15	0.05	0.36	0.78
7	0.50	−0.50	−0.05	0.45	0.29
9	0.28	−0.21	0.05	0.43	0.69
10	0.23	−0.50	0.03	0.32	0.60
8	0.00	0.00	0.00	1.00	0.00

the covariance or correlation matrix is used as input for the analysis, since they differ only by the rescaling of each variable. The data in this study were deemed to be suitably normally distributed, and the large sample size allowed significance tests of hypotheses on the number of factors. On the initial analyses three factors were found inadequate to account for the observed correlation matrix; and an attempt to fit four factors resulted in an improper solution, with variable 8 having zero uniqueness. Thus three factors were fitted to the correlation matrix with variable 8 'partialled out', as previously described. The results of this analysis are shown in Table 4.5, if the last line and fourth column are ignored. The table shows the loadings of the (now nine) variables on the three factors, and the associated uniqueness. The test showed that this was an acceptable solution to the subanalysis. Thus variable 8 was reincorporated by adding the last row of the table, a zero uniqueness, and its correlations with the variables in the fourth column. The resulting four-dimensional solution may now be treated as though it is a simple analysis of the initial correlation matrix, and the complications of the initial improper solution forgotten.

The loadings matrix of Table 4.5 was submitted to a promax rotation procedure

Table 4.6. Promax rotation of loadings in Table 4.5

Factor correlation matrix

	1.	2.	3.	4.
1.	1.00			
2.	−0.50	1.00		
3.	0.26	−0.53	1.00	
4.	0.74	−0.48	0.21	1.00

Factor variances

1.	2.	3.	4.
1.33	1.51	1.05	1.34

(Hendrickson and White (1966) recommend that promax be applied to loadings that are derived from an orthogonal rotation, and in this example the varimax procedure was used first, as a pre-processor). The resulting obliquely rotated axes define factors whose correlations are shown in Table 4.6. Note that there is a generally high level of correlation between these factors, particularly betwen 1 and 4, but that factor 3 correlated highly only with factor 2. All four factors are roughly equal in importance in determing the overall structure, as is indicated by their variances. In Table 4.7 are given the loadings on the promax rotation (only those values outside 0.20 are shown, for clarity) and the factor structure matrix. It should be remembered that the direction of an axis, in the sense of which end is 'positive' and which 'negative', is arbitrary. Thus by choosing to reverse axis 2 in this example, which involves simply changing the sign of all its correlations and loadings values, all the entries in Table 4.6 and Table 4.7 can be made positive. It is worth remarking that a rotation should not be regarded as a process of *amending* factors, but rather of deriving *different* factors. This is emphasized by the arbitrary shifting of variable 8 from factor 4 to factor 3 in the loadings matrix during rotation.

Thus variable 8 appears to constitute a factor by itself, but one which is correlated with factor 2. This factor is the one that determines three of the other four variables in the lower half of the table. Variable 6 does not load heavily on any

Table 4.7. Promax factor loadings of variables

Vbl	Factor 1.	2.	3.	4.
1	–	–	–	0.56
2	0.75	–	–	–
3	0.69	–	–	–
4	0.47	–	–	–
5	–	–	–	0.94
6	–	–	–	–
7	–	−0.83	–	–
9	–	−0.37	–	–
10	–	−0.73	–	–
8	–	–	0.98	–

Correlations of factors with variables

Vbl	Factor			
1	0.66	−0.50	0.19	0.76
2	0.60	−0.27	0.13	0.35
3	0.82	−0.43	0.19	0.69
4	0.58	−0.27	0.18	0.51
5	0.52	−0.29	0.13	0.80
6	0.30	−0.44	0.35	0.26
7	0.42	−0.84	0.43	0.46
9	0.33	−0.52	0.42	0.28
10	0.19	−0.61	0.30	0.16
8	0.25	−0.55	1.00	0.21

factor, and consequently has a high uniqueness (Table 4.5). It is thus a variable 'by itself', and uncorrelated with the underlying structure of the other nine. The remaining variables group neatly, with 1 and 5 being determined by factor 4, and the rest by factor 1. If the identifying of the first two factors is possibly to be expected, the behaviour of variables 6 and 8 must be considered interesting.

4.4.3 Factor Scores

In determining the score for each individual on the derived factors a new consideration arises. The calculation of component scores in Section 3.1 was a straightforward matter, and these scores could be used either to construct an approximation to the data matrix or be used in their own right in subsequent analysis. The least squares principle employed there gave rise to both an approximate covariance matrix and a set of component scores. The principle behind CFA, however, allows only the covariance matrix to be determined, and the factor scores themselves are indeterminate. In other words there are infinitely many ways in which we can pick the factor scores **F** and error variables **E** that satisfy the variance and correlation conditions (i) and (ii) in Section 4.4, and which also give a decomposition of the data matrix.

$$\mathbf{Y} = \mathbf{LF} + \mathbf{E}$$

This problem of indeterminacy arises in CFA because the p variables are transformed into a greater number $q + p$ of factors and errors; by contrast, in PCA there are never more than p variables of any sort.

The indeterminacy has no possible way of being resolved, but it is possible to find a set of surrogate scores, which will be denoted $\tilde{\mathbf{F}}$, that are roughly similar to all the possible scores **F**. These do not themselves give a breakdown of the data matrix **Y** which meets the requirements of Section 4.4, but are used because all possible factor scores **F** can be derived by arbitrary, but hopefully small, changes in $\tilde{\mathbf{F}}$. The degree of arbitrariness in this procedure (i.e. the discrepancies between **F** and $\tilde{\mathbf{F}}$) depends upon the explanatory power of the individual factors in the analysis; if 'small' or 'weak' factors occur their scores may vary considerably from the surrogates $\tilde{\mathbf{F}}$. This feature of factor score indeterminacy has at times led to strong criticism of CFA. It should be borne in mind, however, that this indeterminacy in no way affects the loadings, uniqueness, or their associated statistics.

The classical approach to the problem of factor scores, as given by Lawley and Maxwell (1971) for instance, is to regard the surrogates $\tilde{\mathbf{F}}$ as estimates of the true factor scores **F** that have properties which are optimal (in a least squares sense) with respect to possible variations in the true errors **E**.

The exact estimation formulae for choosing the surrogates $\tilde{\mathbf{F}}$ need not concern us here, but are given, for instance, in Lawley and Maxwell (1971). Most computer programs will produce scores automatically. However there are two different ways of choosing $\tilde{\mathbf{F}}$, the first due to Thomson (1951) and the second to Bartlett (1938).

Thomson's method, known sometimes as 'the regression method', produces a set of scores that have minimum variability, but a slight bias. Bartlett's method produces unbiased scores at the expense of slightly greater variability. There are no hard and fast rules on which should be used, although the present writer's opinion is that unbiasedness is not a particularly useful property. If the factors are fairly powerful and the error variances not large, then the two methods will not give very different answers, in general.

4.4.4 Applications

In that it fits a precise model, CFA does not have the adaptability to other uses that PCA has. Its most obvious use is that of separating 'real' or 'true' scores from a set of data that is compounded with some sort of measurement error. This means that it may be used in solving the problem of regression analysis when the regressor variables are measured with error. Maxwell (1961) discusses how one might use a CFA of the regressor set as a pre-processing analysis for the regression analysis. It should be noted that used in this way, where interest centres upon linear combinations of the regressor variables, the difficulties of determining a rotation position are played down. However, the difficulties of determining factor scores might limit its applicability in subsequent prediction from the regression formula. Maxwell (1961) also considers the equivalences between CFA and analysis techniques that employ the regressing of each variable on all the others.

It seems likely that, in comparison with PCA, CFA is not as robust an analysis; and as a result there may be some equivocation in deciding whether the smaller factors or subspace dimensions should be used. Because of this, some writers (e.g. Gorsuch, 1974) argue that in all except the most exploratory analysis great care should be taken in the selection of the variables to be used in a CFA. It is often recommended, for example, that some thought be given to the anticipated substructure and four or five 'marker' variables be chosen for each factor. It is obvious that in doing so one is forcing the expected results into the analysis to a certain extent, both in terms of the number and the type of factors obtained. Nonetheless one derives, if nothing else from such an analysis, a measure of the communalities of the variables with the presumed factor subspace.

When being used in a purely confirmatory role, that is, when the entire data set is simply being used as a test of some previously determined hypothesis about the factor structure, an interesting development by Joreskog (1969) is applicable. Termed the 'restricted factor model' by Lawley and Maxwell (1971), who discuss it in some detail, it provides a method of testing simultaneously the number of factors and the factor structure. A model is specified by stating the number of factors, and which variables will have zero loadings on particular factors. Thus the analyst is specifying which variables are independent (in the sense of zero partial correlation) of which factors. By assuming multi-normality, a large sample likelihood ratio test of the overall hypothesis becomes available. It should be noted that this procedure does not fit into the overall schema mentioned at Section 4.2 of first deriving an approximating space and then determining an axis rotation; but similar results (without the test of significance) might be achieved within this format by using a Procrustes rotation on an ordinary unrestriced CFA solution.

4.5 SOME EXTENSIONS

The foregoing is a description of the simplest sort of component and factor analyses. Various extensions have in the past been tried and now fallen into disuse, more or less: for instance the concept of quadratic, or shape, components; and of higher order, or group, factors. Current developments are mainly in two directions: firstly, within the classical linear model; and secondly, into non-parametric analyses. Within the standard model, work has been done in generalizing factor analysis to cover situations where more than one group of individuals are being studied. The Lawley and Maxwell (1971) text considers two populations within each of which the same structural factors exist, but with different correlations amongst themselves. This approach is taken up again by Carroll and Chang (1971) in their Individual Differences Scaling (see Chapter 6). There the case is considered where several groups share the same factors, which have different 'saliences', or variances, within each group. Joreskog (1970) extends the classical linear model still further by incorporating a factor analysis model with a multivariate analysis of variance model into what he terms a general analysis of covariances. These models require variables measured on interval scales so that correlations may be formed. In some cases ordinal data suffices if rank correlation coefficients can be used, as is often the case in simple CFA and PCA. By considering not specifically correlation coefficients, but any coefficient of association, the principles of PCA lead directly into Multi-dimensional Scaling. A vast number of procedures exist in this field (see, for instance, Shepard, 1972), which typically deal with only the ordinal properties of the coefficients in deriving low-dimensional approximations to the observed configurations. These areas are subject to pitfalls, both conceptual and computational, of their own, and are dealt with in Chapter 6 and in Chapter 2 in Volume 2.

REFERENCES

Andrews, D. (1972), 'Plots of high dimensional data', *Biometrics,* **28**, 125–136.

Box, G. E. P. (1949), 'A general distribution theory for a class of likelihood criteria', *Biometrika,* **36**, 317–346.

Bartlett, M. S. (1938), 'Method of estimating mental factors', *Nature (London),* **141**, 609–610.

Bartlett, M. S. (1954), 'A note on the multiplying factor for various χ^2 approximations', *J.R.S.S. (B),* **16**, 296–298.

Carroll, J. B. (1953), 'An analytical procedure for approximating simple structure in factor analysis', *Psychometrika,* **18**, 23–27.

Carroll, J. B. and Chang, C. (1970), 'Analysis of individual differences in multi-dimensional factors', *Psychometrika,* **35**, 283–320.

Cattell, R. B. (1946), *Description and Measurement of Personality*, New York: World Book Co.

Cattell, R. B. (1952), *Factor Analysis,* New York: World Book Co.

Clarke, M. R. B. (1970), 'A rapidly convergent method for maximum likelihood factor analysis', *Brit. J. Math. and Stat. Psychol.,* **23**, 43–52.

Daling, J. R. and Tamura, H. T. (1970), 'Use of orthogonal factors for selection of variables in a regression equation', *J.R.S.S. (C),* 260–268.

Gorsuch, R. L. (1974), *Factor Analysis,* W. B. Saunders Co., Philadelphia.

Gower, J. (1966), 'Some Distance properties of latent root and vector methods used in multivariate analysis', *Biometrika*, **53**, 176–182.

Guttman, L. (1953), 'Image theory for the structure of quantitive variates', *Psychometrika*, **18**, 277–296.

Guttman, L. (1954), 'Some necessary conditions for common factor analysis', *Psychometrika*, **19**, 149–161.

Harman, H. H. (1960), *Modern Factor Analysis*, Chicago: University of Chicago Press.

Harris, C. W. (1962), 'Some Rao–Guttman relationships', *Psychometrika*, **27**, 247–260.

Harris, C. W. and Kaiser, H. F. (1964), 'Oblique factor analytic solutions by orthogonal transformations', *Psychometrika*, **29**, 347–359.

Hawkins, D. M. (1973), 'On the investigation of alternative regressions by principal component analysis', *J.R.S.S. (C)*, 275–286.

Hawkins, D. M. (1974), 'The detection of errors in multivariate data using principal components', *J.A.S.A.*, **69**, 340–345.

Hendrickson, A. E. and White, P. D. (1966), 'Promax: a quick method of rotation to a oblique simple structure. *Brit. J. Math. and Stat. Psychol.*, **17**, 65–70.

Hershon, H. (1975), 'Unpublished research memorandum', *ARU*, London: Inst. of Psychiatry.

Hotelling, H. (1933), 'Analysis of complex statistical variable into principal components', *J. Educ. Psychol.*, **24**, 417–441.

Howe, W. G. (1955), 'Some contributions to factor analysis', *ORNL 1919*, Oak Ridge, Tennessee.

Hurley, J. and Cattell, R. B. (1962), 'The procrusts program: producing a direct rotation to test an hypothesized factor structure', *Behav. Sci.*, **2**, 258–271.

Jolliffe, I. T. (1972), (1973), 'Discarding variables in a principal component analysis', *J.R.S.S. (C).*, **21**, 160–173 and **22**, 21–31.

Joreskog, K. G. (1967), 'Some contributions to maximum likelihood factor analysis', *Psychometrika*, **32**, 443–482.

Joreskog, K. G. (1969), 'A general approach to confirmatory maximum likelihood factor analysis', *Psychometrika*, **34**, 183–202.

Joreskog, K. G. (1973), 'Analysis of covariance structures' *Biometrika*, **60**, 314–330.

Kaiser, H. F. (1958), 'The varimax criterion for analytic rotation in factor analysis', *Psychometrika*, **23**, 187–198.

Kaiser, H. F. (1963), 'Image analysis' in *Problems in Measuring Change*, ed. Harris, Univ. Wisconsin Press, 156–166.

Kaiser, H. F. (1970), 'A second generation Little Jiffy', *Psychometrika*, **35**, 401–410.

Kendall, M. G. and Stuart, A. (1961), *Advanced Theory of Statistics*, London: Griffin.

Lawley, D. N. and Maxwell, A. E. (1963), *Factor Analysis as a Statistical Method*, London: Butterworth.

Lawley, D. N. and Maxwell, A. E. (1971), *Factor Analyis as a Statistical Method*, 2nd edition, London: Butterworth.

Lawley, D. N. and Maxwell, A. E. (1973), 'Regression and factor analysis', *Biometrika*, **60**, 331–338.

Matsakis, J. (1957), 'Sur la classification et les affinités de quelques Isopodes terrestres', *C.R. Acad. Sci. (Paris)*, **245**, 584–586.

Maxwell, A. E. (1961), 'Recent trends in factor analysis', *J.R.S.S. (A)*, **124**, 49–59.

Maxwell, A. E. (1971), 'Estimating true scores and their reliabilities in the case of composite psychological tests', *Brit. J. Math. Stal. Psychol.*, **24**, 195–204.

Newhams, J. O. and Wigley, C. (1954), 'The quartimax method: an analytical approach to orthogonal simple structure', *Brit. J. Stat. Psychol.*, **2**, 81–89.

Rao, C. R. (1955), 'Estimation and tests of significance in factor analysis', *Psychometrika*, **20**, 92–101.

Rao, C. R. (1968), *Linear Statistical Inference*, London: Wiley.

Russell, M. A., Peto, J. and Patel, U. (1974), 'Classification of smoking by factorial structure of motives', *J.R.S.S. (A)*, **137**, 313–346.

Schonemann, P. H. (1966), 'A generalized solution of the orthogonal procrustes problem', *Psychometrika*, **31**, 1–13.

Shepard, R. N. ed. (1972), *Multi-Dimensional Scaling in the Behavioral Sciences*, Vols. I, II, New York: Seminar Press.

Spearman, C. (1904), 'General intelligence objectively determined and measured'. *Am. J. Psychol.*, **15**, 201–293.

Thomson, G. H. (1951), *The Factorial Analysis of Human Ability*, London: London University Press.

Thurstone, L. L. (1947), *Multiple Factor Analysis*, Chicago: University of Chicago Press.

Two of the available computer programs:

Joreskog, K. G. (1967), 'UMFLA – a computer program for unrestricted maximum likelihood factor analysis.' *Res. Memo 66–20*, Princeton, N.J.: Educational Testing Service.

Nie, N., Hull, C., Jenkins, J., Steinbrenner, K., and Bent, D. (1975), *S.P.S.S.* (2nd Edition), ch. 24, McGraw-Hill, New York.

Chapter 5

Latent Structure Models

A. Fielding

5.1 INTRODUCTION

The use of latent structure models has been recommended and appropriate methodology developed by P. F. Lazarsfeld and others over the course of the past twenty-five years. Lazarsfeld and Henry (1968) provide a comprehensive survey of much of this work.

The models are a family of scaling or classification models implicitly measuring unobservable phenomena which cannot be satisfactorily operationally defined. They provide a method of differentiating between people, objects, or collectivities either by classifying, ordering, or positioning them along some continuum with respect to underlying characteristics that cannot be explicitly measured.Examples of such characteristics abound in the behavioural sciences: morale, anti-semitism, intelligence, job satisfaction, alienation from bureaucracy, cultural sophistication, racial prejudice and prudence. Such hypothetical continua are often referred to as *latent spaces*, and the corresponding variables, *latent variables*. However, whilst measurement of such a latent variable as racial prejudice cannot be made, there may be well defined indicator or manifest variables, closely related to it, which are measurable. One can, for instance, observe whether a person approves or disapproves of a particular piece of government immigration legislation, or whether he numbers members of a particular race among his friends and acquaintances.

Many methods have been suggested for such a scaling of individuals on the basis of these observed manifest variables. In common with these methods and their underlying models, latent structure models attempt to provide a structural explanation of the relationships holding between the latent and manifest spaces, and serve as a medium through which individuals can be assigned a position on the latent space on the basis of their observed position in the manifest space. The latent space is a formalization of an underlying concept. However, observations should be made on variables relevant to this concept. In a survey connected with racial prejudice, for instance, a problem on integral calculus, or a question on the acceptability of a certain food (unless it happened to be curry or some such ethnic delicacy) would not normally be included in the analysis. An attitude question on the current government's immigration policy almost certainly would. This point is not trivial in view of the widespread malpractice of throwing every bit of survey information into a black box without regard to the appropriateness or reasonableness of the model governing the mechanism inside the particular black box.

Prior to data analysis, formalizations of the way in which the manifest variables are related to the latent ones are required. Usually this takes the form of the construction of a mathematical model. What must be incorporated if the mathematical model is to be a realistic attempt at explaining social science phenomena is a probabilistic element. Thus a person highly racially prejudiced may be more likely to disapprove of a particular piece of government immigration policy than one more tolerant, but in a particular set of circumstances both may give the same response when asked about it. They may even give answers at variance with their respective relative prejudices. The essential point is that before the event a tolerant person is regarded as having a higher probability of giving a favourable response.

Latent position and manifest position are, then, not in a correspondence connected by a rigid functional relationship as in deterministic models. The population under study is assumed to be distributed over the latent space in some way, and the response of an individual at a particular position in the latent space is assumed to have a probability distribution over the manifest space conditional on that position. Thus, for instance, individuals at a particular level of racial prejudice might have a probability of agreeing with a particular piece of immigration legislation of 0.8, people at other positions, a probability of 0.4. The particular models within this family of probability models are different specifications of the form of this conditional distribution, the type of latent space and the form of the distribution of individuals over that space. In its most general form a latent structure model does subsume many other scaling models. It can be shown, for example, that the commonly used Guttman scaling technique, which takes as its model the existence of a Guttman perfect scale, is a special case.

A simple example of the latent structure model might make its special features and potential uses more evident. Suppose as before one is categorizing a group of people with respect to their racial prejudice. It might be supposed from external or internal considerations that three categories would form a suitable formalization of the latent space, with relative frequencies u_1, u_2, u_3. It is proposed to ask a sample of people a series of p attitude questions thought *a priori* to be indicators of racial prejudice. Each may be of the simple agree/disagree type. Thus the manifest variables x_i, $i = 1, \ldots, p$ can each be treated as taking the value unity if the response is agreement and zero otherwise. A latent structure modelling of the situation would specify that the probability of a particular set of responses is conditional and different for each of the latent categories. To be more specific it might be supposed that conditional on the latent categories the manifest variables are independent.

This assumption, called the *axiom of local independence*, is a crucial one in latent structure models. This latent structure model, then, reduces to a series of parameters $p(x_i = 1 \mid j)$, $i = 1, \ldots, p$, $j = 1, 2, 3$ which are the conditional probabilities of an agreeable response to each item i given the category j in the latent space. The data provided are the response patterns of each sample individual on the p attitude questions. Fitting the model involves estimating these parameters and u_1, u_2, u_3 from the sample data. Assuming the model has been appropriately chosen these parameters then form a basis for the classification of individuals into one of three categories of racial prejudice. This may be an exercise in its own right or a prelude to further analyses of the data which include the latent variables. It will be seen that such a model fitting may also serve as a data summary or reduction method, and may indeed form a framework for exploring hidden patterns of relationships between the manifest variables. These uses and applications will be discussed in the last section.

This example provides a simple type of latent structure model. In the next section the general model will be discussed, together with a general discussion of the problem of specifying parameters when the population distribution on the space of manifest variables is known. The relationship to other scaling models and methods of analysis will be briefly explored. Section 5.3 will discuss the various specializations of the general model that have found wide application in practice, and various methods of finding the parameters for these models from population manifest data. It will be seen in Section 5.4 that though these methods may equally well be applied to sample data, additional considerations may arise which suggest other estimation procedures which may be more appropriate. The general problem of fitting and testing models and its relationship to parameter identification are also explored there.

5.2 THE GENERAL MODEL AND ITS RELATIONSHIP TO OTHER METHODS

A general approach to the formulation of a probabilistic scaling model may be stated as follows (cf. Anderson, 1959). Let $x_1, x_2, \ldots, x_p$ be a series of manifest variables. In principle, each could be at any level of measurement. Further let y_1, $y_2, \ldots, y_m$ be latent variables which again could be specified at any level of measurement. The number of latent variables m is usually much less than p since a latent structure model may be viewed as a data reduction medium which reduces the full set of manifest observations to a few latent values. Indeed in the context of most of the models to be discussed later m will be equal to one. It is assumed that $x_1, x_2, \ldots, x_p$ have a joint probability distribution conditional on $y_1, y_2, \ldots, y_m$ denoted by

$$\phi(x_1, x_2, x_3, \ldots, x_p \mid y_1, y_2, \ldots, y_m)$$

If the x_i are all continuous variables, ϕ is a density function, but if all are discrete it is a set of probabilities. Similarly, let the distribution of the latent variables be denoted by a density or probability

$$h(y_1, y_2, y_3, \ldots, y_m)$$

From the rules of probability the unconditional marginal distribution of x_1, $x_2, \ldots, x_p$ is

$$f(x_1, x_2, \ldots x_p) = \int \phi(x_1, x_2, \ldots, x_p \mid y_1, y_2, \ldots, y_m) \, h(y_1, y_2, \ldots, y_m) \mathrm{d}y_1 \, \mathrm{d}y_2 \ldots \mathrm{d}y_m. \tag{1}$$

The function f can again be interpreted as a density or probability (with an appropriate change of integration to summation if the y are discrete).

Further, by a straightforward application of Bayes' theorem, given an individual's set of values on the manifest variables $X_1, X_2, X_3, \ldots X_p$ the density or probability of the individual's y scores over the latent space is

$$g(y_1, y_2, \ldots, y_m \mid X_1, X_2, \ldots, X_p) = \frac{\phi(X_1, X_2, \ldots, X_p \mid y_1, y_2, \ldots, y_m) h(y_1, y_2, \ldots, y_m)}{f(X_1, X_2, \ldots, X_p)} \tag{2}$$

A position in the latent space could be assigned to an individual from (2) given his set of values X_1, X_2 $X_3, \ldots, X_p$. The simplest allocation method is a maximum likelihood approach where an individual is given his most likely position; that is the point Y_1, $Y_2, \ldots, Y_m$ which maximizes (2). Alternatively, different weights may be attached to different positions if it is considered that misclassification is more serious in certain positions than in others. This involves a decision rule involving some suitable loss function.

A number of immediate problems with the application of the technique are apparent. The first concerns the nature of the latent variables, their number and their form, which involves the specification of an appropriate model at the outset which may be based on substantive considerations. Thus, in the example we have supposed that an appropriate form for racial prejudice is a three class categorization. Most models to be considered here will involve a single y, although extensions to multidimensional latent space have been suggested. However, a decision on whether this latent variable should be conceived of as continuous, or nominal or ordered categorization, or even a ranking on individuals must be determined from the context of the problem under study. The appropriateness of different choices of latent space can be viewed as a partly practical problem but can sometimes be tested by the manifest data themselves. This is particularly true in the case of deciding on the number of categories for y if the so called Latent Class Model for categorized y is used.

A more important problem is that the use of equation (2) requires that ϕ, h, and f are known. Indeed, if this were the case the analysis and scaling would be trivial. Now the function f is on an instrumentally different standing from the other two. If information is available on manifest variables for the population under study then f will be known. The usual situation is that ϕ and h are not known, they cannot be inferred from f uniquely, and if any further progress is to be made some additional assumptions about their form are required.

The fairly general assumption of latent structure models is the axiom of local independence which states that given a position on the latent space the manifest variables are independently distributed. The function ϕ with one latent variable can then be written

$$\phi(x_1, x_2, \ldots, x_p \mid y) = \phi_1(x_1 \mid y)\phi_2(x_2 \mid y) \ldots \phi_p(x_p \mid y). \tag{3}$$

Although it seems that this is a restrictive form of ϕ in the general scaling model there are substantive reasons for making this assumption in the behavioural sciences. The basic postulates are that sets of individuals with specific latent characteristics are similar, that the latent characteristics are the important ones, and that once these have been determined behaviour is random. This implies that responses to the manifest items conditional on a specific position in the latent space will be independent. In any event, a survey analyst will normally only use manifest variables because they are related and partially express the underlying construct in the way it has been conceived and hence he will enforce local random behaviour. In the population as a whole one would expect a high degree of interdependence between the manifest observations. Latent structure models by use of the axiom are designed to account for this interdependence. They postulate that in general terms this interdependence is due to the common dependence on the latent variable. In a way this is not unlike saying that in a general linear model all higher order generalized correlations controlling for the same variable (i.e. the equivalent of the latent one) at zero.

The functions $\phi_i(x_i \mid y)$ being the conditional densities or probabilities of the manifest variables are called the *trace functions*. It is these together with the latent distribution h that one wishes to specify in solving the 'structure', in order to see how the manifest variables depend on the latent ones and to use the equation (2) as an aid to scaling individuals, and estimate parameters of the functions. In the language of structural relationships (c.f. Koopmans and Reiersøl, 1950) one is faced with an *identification problem*. Given equation (3) expressing a general model of local independence, and knowing f, how are the trace functions and latent distribution to be specified uniquely? In other words, can the structure be identified? It may be apparent that the general model subsumes many different structures which could be consistent with and generate the same manifest distribution f. The local independence axiom will not on its own guarantee identification. Therefore additional assumptions about the form of the trace functions, latent distribution or both will be required to identify the structure. The general local independence model together with particular sets of assumptions about the nature of these functions generate subfamilies of latent structure models.

The *Latent Class Model* of Section 5.3.1 assumes that the latent variable is a classification with m categories, and the manifest variables are dichotomies. Here the trace functions are simply binomial probabilities and the latent distribution is specified by a set of probabilities $u_1, u_2, \ldots, u_m$ summing to unity. In Section 5.3.2 the *Latent Profile Model* assumes a similar latent categorization but the manifest variables are continuous. No particular distributional form is assumed for these variables but interest centres on the specification of the first moments (e.g. means and variances) of the trace functions. Manifest dichotomies and a continuous latent variable are discussed in the following section and various types of trace function are considered. *The Latent Content Model*, for instance, assumes that

$$P(x_i = 1 \mid y) = a_i + b_i y^{d_i}.$$

Two common distributional forms for $h(y)$, the Uniform and Beta, are explored.

Here specification of the structure and solving the model involves finding a_i, b_i, d_i, for each manifest item, and the parameters of $h(y)$.A series of other models is discussed in the last part of Section 3 including the *Latent Distance Model* where it is desired to impose an ordering of categories only on the latent space.

Even within these subfamilies it is possible that different structures are compatible with a known manifest distribution f, corresponding to different sets of parameter values. Much of the early work on the different types of latent structure model concerned itself with this problem and was directed to discovering under what conditions the structure, i.e. trace functions and latent distribution, could be identified uniquely. To be of any real use it is clear that a submodel must be such that a particular structure within it is capable of identification from a knowledge of f. However, although in many situations a particular submodel does not entirely guarantee identifiability of structure it may be possible to specify uniquely certain aspects of the structure which are sufficient for all practical purposes. It may not be possible, for instance, to identify $\phi_i(x_i \mid y)$ for a given model, although one may be able to solve uniquely for $E(x_i \mid y)$, the mean value of x_i given y.

A particular example may make the problem of identifying unique structures clearer. One way of specifying a submodel, for instance, is to assume that the parametric forms of the trace functions and latent distributions are known. They could, for instance, be assumed normal with unknown means and variances. Here $\phi_i(x_i \mid y, \theta_{1i}, \theta_{2i}, \ldots\ldots, \theta_{q_i})$, $h(y \mid \psi_1, \psi_2, \ldots, \psi_r)$ are assumed known functional forms but with parameter values denoted by the θ's and ψ's. This would form a submodel within the general latent model and all the possible combinations of sets of values of the θ's and ψ's within the parameter space could be particular structures. The problem of identification becomes one of specifying a particular set of values of these parameters uniquely from the manifest distribution. There may, of course, be circumstances in which the structure is only partly identifiable in the sense of the last paragraph, in which case only some of the parameter values such as moments or functions of them can be specified.

The identification problem has been a subject of much enquiry in the field of econometrics (e.g. Marschack, 1972; Koopmans, 1949). So far, of course, the problem of estimation when information on f is available only from a sample has not been considered. The identification problem is assumed to relate to known population forms. Estimation problems obviously do arise and will be discussed in Section 5.4. However, it is comforting to know that if the structure is specified in parametric form and consistent estimates of the parameters can be found then they must be identifiable in the population (McHugh, 1956).

Latent structure models thus provide a general framework which, subject to the solution of identification and estimation problems, enable one to study the interrelationships of indicators and their connection with underlying concepts. The well known factor analysis model (see Chapter 4) purports to do the same thing.

5.2.1 Relationship to factor analysis

It might appear that factor analysis and latent structure analysis are derived from different conceptual models. However, in its least restrictive form the factor analysis model is a particularization of equation (1), and with usual assumptions falls within the domain of a multidimensional extension of equation (3). The basic assumption of factor models is that each manifest quantitative variable x_i can be expressed as a linear combination of a number of quantitative latent variables plus a unique or error term:

$$x_i = a_{1i}y_1 + a_{2i}y_2 \ldots a_{mi}y_m + \epsilon_i \quad i = 1, 2, \ldots, p \tag{4}$$

Since the y_i's and ϵ_i's are random variables this is just a specification of model (1). If, as is usually the case, the ϵ_i's are assumed independent of each other then the local independence axiom holds, and we have a latent structure model with a continuous and multidimensional latent space. With additional assumptions such as multivariate normality of the y's and normality of the e_i's independent of the y's, the strictest model of factor analysis ensues. $E(\epsilon_i) = 0$ implies in the case of equation (4) that the specification of the model has a trace function with regression $E(x_i | y)$, a linear function of the latent variables. Incidentally if multivariate normality is assumed then the local independence assumption is automatically satisfied by specifying zero correlations between the ϵ_i's. Thus equations in higher order dependencies, such as third order moments of continuous manifest variables or third order joint proportions, which will usually be required to solve other latent structures, impose no additional restrictions and are not required. Thus the usual specification of factor analysis in terms of analysis of the correlation matrix is perfectly compatible with solving a latent structure model.

The confusion between the two approaches, which usually arises in practice, may be due to the fact that latent structure analysis has been largely concerned with a single latent variable and more often one which is a classification. Thus the practical latent structure analyses use models which are in some senses more general and in others more restrictive than factor analysis. Yet the use of non-linear regression functions, such as that for the latent content model discussed previously, indicates how a restrictive factor analysis model may be generalized to take care of situations where a non-linear relationship between latent and manifest variables may be thought to be more appropriate (Green, 1952; Gibson, 1960). The multidimensionality of the latent space could possibly be explicitly considered, but, apart from the obvious example of the particularization to a linear factor model, has not so far been investigated. The model with latent classes and quantitative manifest variables, which has been called the Latent Profile Model, could be thought of as a partitioning into an ordered cross-classification of a continuous latent space of several dimensions. If linearity of the regression function within classes was also assumed, and a partition specified, then clearly this model and linear factor analysis would be mutually complementary. The assumptions of both would be satisfied and similar conclusions about the structure ought to be reached. However, the factor analysis results might be more appropriate for telling one about the way the manifest variables are related, whereas with latent groups one might have a rational way of classifying individuals under study (Mardberg, 1973). Additionally, latent profile models do not in principle require linear trace regressions, even if a continuous latent variable was assumed to underly the grouping. It appears that the general class of latent structure models provides pointers to a more realistic analysis of the type of data to which a routine linear factor analysis has been normally applied.

5.3. TYPES OF LATENT STRUCTURE MODEL

The general latent structure model (3) is specialized by further assumptions or by restrictions on the form of the trace functions. For the moment the case of just one latent variable is considered. At one extreme is the model where all the manifest and latent variables are categorized. Here the problem of specifying functions does not arise since it is a multinomial situation and parameters are cell probabilities. At the other extreme all variables are continuous. Within these limits there are many possible variants.

5.3.1 The Latent Class Model

Here the assumption is that there are m latent classes ($m > 1$) and that the manifest variables are dichotomous attributes or items. Dichotomous attributes are common in many surveys, particularly those concerned with opinions or attitudes where questions demanding a 'yes' or 'no' answer are frequently posed. To solve the structure requires a solution of the distribution $u_1, u_2, u_3, \ldots, u_m$ of individuals over latent categories, and the trace functions in this case are $\pi_i^\alpha (i = 1, \ldots, p, \alpha = 1, 2, \ldots, m)$, the probabilities of an individual in latent class α giving a positive response to each manifest variable i. There are a total of $mp + m = m\ (p + 1)$ parameters. The local independence axiom states that for a given latent class the probability of an individual having a particular combination of responses for any subset of items is equal to the product of the marginal probabilities of giving those responses to the items. Thus if π_{ijk}^1 is the joint probability of giving a positive response to all items i, j and k in latent class 1, then this would be equal to $\pi_i^1 \pi_j^1 \pi_k^1$, the product of the marginal probabilities of giving a positive response to each of the items within class 1. As a matter of notational convenience a subscript such as i or j denotes a positive response to the corresponding variable whilst a bar over the letter as in $\bar{i}$ or $\bar{j}$ denotes a negative response. Thus $\pi_{i\bar{j}}^1$, would denote the joint probability of giving a positive response to variable i, and a negative one to j in latent class 1. π_{ij} without a superscript denotes the manifest proportion having that response pattern, and corresponds to parameters of the f distribution in the general model. These are assumed known (or, as will be seen later, estimable). Two basic sets of relationships must hold between the parameters:

$$u_1 + u_2 \ldots + u_m = 1$$

$$\pi_i = \pi_i^1 u_i + \pi_i^2 u_2 \ldots + \pi_i^m u_m \quad (i = 1, 2, \ldots, p)$$

The first states that the sum of latent probabilities must equal one; the second that the manifest probability for each item in the overall population giving a positive response to item i is the weighted sum of the latent class probabilities, and is referred to as the first order probability equation. Local independence implies relationships between second order probabilities,

$$\pi_{ij} = \sum_{\alpha=1}^{m} u_\alpha \pi_{ij}^\alpha = \sum_{\alpha=1}^{m} u_\alpha \pi_i^\alpha \pi_j^\alpha \quad (i, j = 1, \ldots, p, i \neq j)$$

and third order relationships,

$$\pi_{ijk} = \sum_{\alpha=1}^{m} u_\alpha \pi_i^\alpha \pi_j^\alpha \pi_k^\alpha \quad (i, j, k = 1, \ldots, m, i \neq j \neq k)$$

with obvious extensions to higher orders. From elementary probability considerations it can be seen that there are $2^p - 1$ such equations linking the observed manifest probabilities to the latent parameters, to give a total of 2^p equations. For an identifiable structure these equations, referred to as the *accounting equations*, will be uniquely solvable.

As a matter of fact the accounting equations could have been set up in a form more in keeping with equation (1) where the $2^p - 1$ probabilities of all the different response patterns, including combinations of positive and negative ones, could be expressed in terms of the latent parameters. However, the two sets of equations are algebraically identical.

One essential requirement for identifiability of this structure is that the number of equations must be at least as great as the number of unknown latent parameters, i.e. $2^p \geqslant m(p + 1)$. It will usually be satisfied if the number of manifest variables is sufficiently large. Now clearly if the structure is identifiable and there are more equations than parameters, it should be possible to find a solution by selecting among some of the equations.

If there is a sufficient number of manifest variables a solution should be possible by selecting equations involving only lower order joint probabilities. An alternative is to use fewer manifest variables and higher order probabilities. The equations selected may of course be constrained by problems of identifiability, since certain selections may not lead to a unique solution. An algebraic solution has been developed using a selection of up to third order equations only (Anderson, 1959), which also specifies conditions for identifiability. Another approach (Gibson, 1955) uses fewer variables but higher order equations. The Anderson solution requires selection of $2m - 1$ manifest variables to form two distinct sets of $m - 1$ each and an additional one called the *stratifier* variable. Since numbering of manifest variables is arbitrary it can be supposed without loss of generality that the first $(m - 1)$ comprise the first selected set and the next $(m - 1)$ the second selected set. The so called additional stratifier variable is given the label k. A selection of the accounting equations which satisfy certain conditions are cast into matrix form and the solution ensues.

*Let **N** be an $m \times m$ diagonal matrix with $u_1, u_2, \ldots, u_m$ along the main diagonal, and zeros elsewhere. Let **D** be an $m \times m$ diagonal matrix with the latent probabilities corresponding to the selected stratifier variable k, $\pi_k^1, \pi_k^2, \pi_k^3, \ldots, \pi_k^m$ along the main diagonal and zeros elsewhere. The $m \times (2m - 1)$ matrix Λ is formed with latent probabilities of the selected sets of items as its entries.

$$\Lambda = \begin{bmatrix} 1 & \pi_1^1 & \pi_2^1 & \cdots & \pi_{2m-2}^1 \\ 1 & \pi_1^2 & \pi_2^2 & \cdots & \pi_{2m-2}^2 \\ 1 & \pi_1^3 & & \cdots & \\ 1 & & & & \\ \cdot & & & & \\ \cdot & & & & \\ \cdot & & & & \\ 1 & \pi_1^m & & \cdots & \pi_{2m-2}^m \end{bmatrix}$$

Let Λ_1 be the $m \times m$ matrix consisting of the first m columns of Λ, and Λ_2 the matrix consisting of the first column of Λ and columns $m + 1$ to $2m - 2$. These will be matrices whose columns are latent probabilities of each of the two selected sets and each bordered by a first column of ones. The matrices Π_1 and Π_2 are formed to consist of certain of the known manifest probabilities up to the third order involving the selected items and the stratifier:

$$\Pi_1 = \begin{bmatrix} 1 & \pi_m & \cdots & \pi_{2m-2} \\ \pi_1 & \pi_{1,m} & \cdots & \pi_{1,2m-2} \\ \pi_2 & \pi_{2,m} & & \pi_{2,2m-2} \\ \pi_3 & \pi_{3,m} & & \pi_{3,2m-2} \\ \cdot & \cdot & & \cdot \\ \cdot & \cdot & & \cdot \\ \cdot & \cdot & & \cdot \\ \cdot & \cdot & & \cdot \\ \cdot & \cdot & & \cdot \\ \pi_{m-1} & \pi_{m-1,m} & & \pi_{m-1,2m-2} \end{bmatrix}_{m \times m}$$

$$\Pi_2 = \begin{bmatrix} \pi_k & \pi_{m,k} & \cdots & \pi_{2m-2,k} \\ \pi_{1,k} & \pi_{1,m,k} & \cdots & \pi_{1,2m-2,k} \\ \pi_{2,k} & \pi_{2,m,k} & \cdots & \pi_{2,2m-2,k} \\ \cdot & \cdot & & \cdot \\ \cdot & \cdot & & \cdot \\ \cdot & \cdot & & \cdot \\ \pi_{m-1,k} & \pi_{m-1,m,k} & \cdots & \pi_{m-1,2m-2,k} \end{bmatrix}_{m \times m}$$

It will be seen that certain of the accounting equations up to the third order can be expressed using these matrices as

$$\Pi_1 = \Lambda_1' N \Lambda_2$$

and

$$\Pi_2 = \Lambda_1' N D \Lambda_2 .$$

Next one considers the matrix

$$\begin{aligned}\Pi_2 \Pi_1^{-1} &= \Lambda_1' N D \Lambda_2 \Lambda_2^{-1} N^{-1} (\Lambda_1')^{-1} \\ &= \Lambda_1' N D N^{-1} (\Lambda_1')^{-1} \\ &= \Lambda_1' D (\Lambda_1')^{-1}\end{aligned}$$

From this it is seen that $\mathbf{D}$ is the diagonal matrix of eigenvalues of $\Pi_2 \Pi_1^{-1}$ and hence the latent parameters $\pi_k^1, \pi_k^2, \ldots, \pi_k^m$ can be found by standard matrix methods. Λ_1' is of course the eigenvector matrix of $\Pi_2 \Pi_1^{-1}$, with each column determined uniquely up to multiplication by a constant if the eigenvalues are all different. Thus if $\mathbf{X}$ is a matrix of eigenvectors

$$\Lambda_1' = \mathbf{X}^{-1} \mathbf{E}$$

with $\mathbf{E}$ diagonal and non-singular. However, the first row of Λ_1' must contain unities, and the diagonal elements of $\mathbf{E}$ must be the reciprocals of the elements of the first row of $\mathbf{X}^{-1}$. Hence Λ_1' can be found and gives the latent probabilities of the first set of selected variables.

Consider now $(\Lambda_1')^{-1} \Pi_1 = (\Lambda_1')^{-1} \Lambda_1' N \Lambda_2 = N \Lambda_2$. The left hand side $(\Lambda_1')^{-1} \Pi_1 = \mathbf{A}$ say is known and also the elements of the first column of Λ_2' are known to be unity. The elements of the diagonal $\mathbf{N}$, the latent class distribution are the elements of the first column of $\mathbf{\Lambda}$. Finally, the latent parameters for the second set of items are known from premultiplying both sides of the above by $\mathbf{N}^{-1}$ to get

$$\Lambda_2 = N^{-1} (\Lambda_1')^{-1} \Pi_1 .$$

A solution is found so far only for the latent parameters of the $2m - 1$ selected variables. However, a matrix Π_3 similar to Π_1 but not necessarily square can be formed. This has the same first column as Π_1, and elements in the first row π_{2m}, π_{2m+1}, etc. In the body of the matrix are the second order joint manifest probabilities π_{ij} ($i = 1, 2, \ldots m - 1; j = 2m, 2m + 1, \ldots,$ etc.). Similarly, a matrix Λ_3 can be found with m rows, where each column is the set of trace probabilities for the manifest items $2m, 2m + 1, 2m + 2, \ldots,$ etc., bordered by a first column of unities. From the second order accounting equations

$$\Pi_3 = \Lambda_1' N \Lambda_3$$

and therefore

$$\Lambda_3 = N^{-1} (\Lambda_1')^{-1} \Pi_3$$

giving the required parameters for the rest of the manifest items.*

The above is an approach to a general solution and if certain conditions are satisfied it will give a unique solution. If the population joint manifest distribution were known and the model appropriate, this solution would also satisfy any higher order equation expressing the population structure. However, in the outline of the solution several conditions necessary for identification have implicitly emerged. Π_1 must have an inverse so we require Λ_1, $\mathbf{N}$, Λ_2 to be non-singular. Thus the diagonal elements of $\mathbf{N}$ or the latent marginals must all be non-zero. This is reasonable since if one or more of the latent classes are empty, the implication is a model misspecification and fewer than m classes should be postulated. For Λ_1, Λ_2 to be non-singular the set of manifest variables must contain at least two distinct subsets with this property. Lastly there must be at least one item, denoted here as k, such that the probabilities of positive response are not equal for any pair of latent classes. If this is the case the eigenvalues of the designated matrix will be distinct. The implication is that the classes are sufficiently discriminatory on at least one item.

This so called basic solution also requires that p the number of variables is greater than $2m - 1$. This solution is, however, only one of many variants of algebraic solutions which use different criteria for selecting among the equations before proceeding to the development of a solution algorithm. Some use fewer than $2m - 1$ manifest items and higher order equations. Others attempt to use the full set of manifest variables. One fruitful approach is to use what have been referred to as ascending matrices. Instead of using just data up to second order joint probabilities of manifest variables, those of higher order and the associated latent parameters replace the equations and matrices used in the basic solution. Once this reformulation is made the solution proceeds in a similar fashion (Madansky, 1960). Here fewer manifest variables are required to generate the first part of the solution. The conditions for identifiability are similar, and as before can give some idea of the number of latent classes m to be chosen. Indeed these conditions imply that any square $m \times m$ submatrix of the second order joint manifest probabilities must be singular. A composite stratifier has also been suggested as an aid to solution (Gibson (1955)). Here the matrix $\mathbf{D}$ does not refer to a single manifest stratifier variable k, but each diagonal entry is a sum (possibly weighted) of the latent probabilities of several stratifier variables. This has been suggested as a way of getting a better separation of the roots of the characteristic equation needed to be solved in the algebraic solutions, and hence ensuring identifiability. The other matrices in the basic solution are of course modified in the appropriate way. In addition Gibson suggests an algorithm for bringing together the steps in the solution into one iteration providing a possibly more efficient computing routine. Algebraic methods involving all the items have been developed using an analogy with factor analysis methods, and involving equations up to the second order (Gibson, 1951). These have been extended to remove an element of rotational indeterminacy inherent in factor analysis methods by use of the third order equations (Green, 1951). These variants are just different ways of selecting equations to use in the algebraic solutions. However, it is in the context of estimation from sample data that real comparisons between them can be made. These will be taken up in Section 5.4.

5.3.2 Latent Profile Models

In Latent Profile Models m classes again represent the latent space, but here the manifest variables $x_1, x_2, x_3, \ldots, x_p$ are assumed continuous. No specific form for the conditional distributions of each x_i is assumed, but some of the conditional moments should be specified from the manifest data. Theoretically, of course, with enough of these moments the distributions could be specified uniquely, but interest initially centres on the conditional means, or regression functions $E(x_i$ latent class $\alpha)$, and the second order central moments or variances. The path of the means for each manifest variable is referred to as its latent profile. The profiles together with conditional variances will constitute a part specification of the trace function in the sense discussed in Section 5.2. The latter moments, together with the latent frequencies $u_1, u_2, \ldots, u_m$ have to be determined from the known (assumed) joint manifest distribution of the x_i for a structure to be identifiable. If the joint manifest distribution is known, as will be assumed, the means μ_i, variances σ_{ii}, covariances σ_{ij} and all central moments of any order for the manifest variables are determined. There are, of course, well known relationships existing between moments.

The accounting equations can be set up using relationships between the manifest moments and latent parameters. For example, the second order equations are

$$E(x_i x_j) = \sigma_{ij} + \mu_i \mu_j = \Sigma u_\alpha \mu_i^\alpha \mu_j^\alpha$$

where $E(x_i x_j)$ is the joint manifest moment of x_i and x_j about zero, σ_{ij} is the manifest covariance, μ_i and μ_j the manifest means for x_i and x_j and μ_i^α, μ_j^α are the conditional means (i.e. latent profile parameters). Similar equations can be written down for higher order moments $E(x_i x_j \ldots x_k)$. These equations have the same structure as the accounting equations for latent class models. Indeed, the latter are just a special case with variables x_i taking 0, 1 with the probabilities π_i^α being μ_i^α. The algebra of the accounting equations carries over either in the form above or with central moments on the left hand side and with the deviations from the manifest means $(\mu_i^\alpha - \mu_i)$ instead of μ_i^α on the right hand side. The conditions for identifiability of the latent class model carry over naturally to the latent profile model. So far only one part of the latent structure, the latent profile means, has been specified. However, other parameters specifying higher order moments of the conditional latent distribution such as the within class variances σ_{ij}^α may be of interest. If the latent distribution is assumed normal, means and variances identify it completely. Accounting equations can be set up in the usual way.

It has been shown that the conditions for identifiability are exactly the same as those necessary for the existence of a basic solution for the latent profile means (Lazarsfeld and Henry, 1968) given earlier in the context of latent class models.

The specification of within class variances, as well as latent profile means, should probably be sufficient for most practical purposes.

5.3.3 Models with continuous latent space

For continuous latent spaces the simplest situation is where the manifest variables are dichotomies and the trace functions are probabilities of x_i taking a positive response expressed as a function of the continuous latent variable y, i.e. $P(x_i = 1 \mid y) = \pi_i(y)$. Thus in the example given in the introduction it may be thought that a more appropriate form for racial prejudice is a continuum. Solving a model requires finding the parameters of the trace line, and possibly a part specification of parameters of the distribution of y. The simplest examples are linear trace lines analogous to those of factor models, i.e.

$$\pi_i(y) = a_i + b_i y$$

Of course, depending on the nature of the assumed distribution of y and its range of values, restrictions might be placed on possible values of a_i, b_i to ensure that $\pi_i(y)$, a probability, lies between zero and one.

In general there is a wide range of possibilities for specification of $h(y)$, the distribution of y, and the form of $\pi_i(y)$ subject only to the constraint $0 \leqslant \pi_i(y) \leqslant 1$.

One form for the trace line functions gives rise to what has been called the *Latent Content Model*:

$$\pi_i(y) = a_i + b_i y^{d_i}$$

with the latent variable y scaled so as to take on values in the interval [0,1]. There are obvious restrictions on the parameters to ensure that this function represents a probability: $0 < a_i < 1$; $0 < a_i + b_i < 1$ and $d_i \geqslant 0$ to ensure that $\pi_i(y)$ is less than or equal to unity in the vicinity of $y = 0$. Also the type of manifest data under study has often been chosen such that $\pi_i(y)$ is realistically supposed to increase with increasing y, so that $b_i \geqslant 0$, although this is clearly not a universally necessary assumption. Varying the parameters gives a wide variety of convex, concave or strictly linear functions.

Moment equations similar to those developed for the latent profile model can be established from local independence considerations. Thus, for instance, if the joint manifest probabilities of a pair of x_i, x_j are taken and the general model of Section 5.2 directly invoked,

$$\pi_{ij} = \int_0^1 \pi_i(y)\pi_j(y)h(y)dy \quad (i, j = 1, \ldots, p, i \neq j)$$

$$= \int_0^1 (a_i + b_i y_i^{d_i})(a_j + b_j y^{d_j})h(y)dy$$

$$= a_i a_j + a_j b_i E(y^{d_i}) + b_j a_i E(y^{d_j}) + b_i b_j E(y^{d_i + d_j})$$

It has been pointed out that without some specification of the distribution of y, or at least some of its moments, not much further progress can be made (Lazarsfeld and Henry, 1968). Two possibilities are the Uniform distribution and a Beta distribution of the first kind. The moments of the Beta of any order are specified by two parameters, and in the Uniform, $E(y^r) = 1/(r + 1)$. Lazarsfeld has considered various algebraic solutions to the latent content model for uniform y, and derived some conditions of identifiability (Lazarsfeld and Henry, 1968). Some special cases are also investigated and an interesting application of Somers (Somers, 1961) from a study of social rigidity is developed.

Another obvious possibility for the latent probabilities is that they are polynomial functions of the latent variable,

$$\pi_i(y) = a_{0i} + a_{1i}y \ldots + a_{ri}y^r.$$

This equation, by suitable restrictions of the coefficients, approximates a wide variety of functional forms, and it allows a wide flexibility in the specification of $\pi_i(y)$. Accounting equations in the fom of moment equations can again be formulated but as before matrix expression facilitates their investigation. These equations are similar to but more complicated than those used in the Latent Class Model.

The special cases of linear and quadratic functions have been considered in some detail by Lazarsfeld and Henry (1968). Polynomial trace line models offer possibilities of a much more flexible nature than standard factor analysis models with their implicit linearities. A general approach for analysing non-linearities using models involving linear combinations of orthogonal polynomials in one latent variable has been considered by McDonald (1962).

5.3.4 Other models

The models presented in the previous sections have been the most widely used in practice, but other assumptions generate further submodels. Perhaps the most general is where manifest and latent variables are all continuous variables. Location and scale values for the latent variable are arbitrary, and without loss of generality it may be assumed that $\mu_y = 0$, $\sigma_y = 1$. As before a restriction may also be placed on the regression function of x_i on y, $E(x_i \mid y)$, with a linear form $a_i + b_i y$ the simplest. From this the relations $\mu_i = a_i$ and $\sigma_{ij} = b_i b_j$ can be derived from local independence considerations and a solution derived analogous to those for dichotomous x_i and linear trace lines. It is also the single factor model of ordinary factor analysis in its least restrictive form. It may also be possible to investigate other aspects of the distribution of y and the conditional variances of x_i as a function of y. It is apparent that other forms for the regression function $E(x_i \mid y)$ might be more appropriate models, and that more flexible relationships than those implicit in ordinary factor analysis could be modelled (McDonald, 1962). In principle the latent space could be made multidimensional and with appropriate assumptions restricting the general model, a basis for a more general approach to non-linear factor analysis established. Work in this area is as yet in its formative stages.

One well known deterministic scaling method is *Guttman scaling.* Here the manifest variables are dichotomous and a continous latent continuum is assumed. $\pi_i(y)$ is assumed to be either zero or one, with all respondents up to some point on the latent space, y_i say, always giving a negative response to item i and after that a positive one. The ordering of the values of y impose an order on the items in the sense that if $y_i < y_j$, then any individual responding negatively to item i will always respond negatively to item j. This technique postulates the existence of $p + 1$ ordered latent class demarcated by the cut-off points $y_i (i = 1, \ldots, p)$. Thus a particular response pattern determines uniquely which of the ordered classes an individual belongs to. In practice a perfect scale of this nature will not exist because of the lack of a probabilistic element as noted before. The *latent distance model* between them, so without loss of generality one can apply the *probability integral*

Here the trace lines take on the special form

$$\pi_i(y) = a_i - b_i \quad y \leqslant y_i$$
$$= a_i + b_i \quad y > y_i$$

The trace lines have three parameters a_i, b_i and the break point y_i. The shape of the density function of y has been left unspecified, but it is clear that the shape of the trace lines is unaffected by a monotonic transformation of y as are the relationships between them, so without loss of generality one can apply the *probability integral transform* and assume that y is uniform over the interval $[0,1]$. Since numbering of manifest variables is arbitrary it can be also assumed that $y_1 < y_2 < y_p$. The analogue to Guttman scaling is that if an individual has the lower probability of responding positively for item j he will also have the lower probability for items i if $i < j$. Here y_i also represents the proportion of the population who have the lower probability of responding to item i.

It can be shown that

$$\pi_i = \int_0^{y_i} (a_i - b_i)dy + \int_{y_i}^{1} (a_i + b_i)dy$$
$$= (a_i - b_i)y_i + (a_i + b_i)(1 - y_i);$$

similarly

$$\pi_{ij} = (a_i - b_i)(a_j - b_j)y_i + (a_i + b_i)(a_j + b_j)(y_j \cdot y_i) + (a_i + b_i)(a_j + b_j)y_j$$

for i, j such that $y_i < y_j$

with similar expressions for higher order manifest responses. Algebraic solutions which include a way of deciding on the appropriate ordering of the items have been developed by Hays and Borgatta (1954), who also give a useful comparison of the latent distance model and a more restricted version in which $a_i = 0.5$, together with some interesting applications. Another illustration of its use in a study designed to measure behaviour in a role conflict situation is reported in Lazarsfeld and Henry (1968).

This generalization of Guttman scaling represents an approach to imposing an ordering on a series of latent classes. The general m-class latent structure model discussed in Section 5.3.1 defines as its stands only a nominal typology. It may be possible to impose an ordering $1, 2, \ldots, m$ if it turns out that $\pi_i^1 < \pi_i^2 < \ldots . . < \pi_i^{m-1} < \Pi_i^m$ for all items i, but this is not usually the case either because the manifest variable parameters are not monotonically related to each other in this simple way, or perhaps because the underlying latent space is not unidimensional so that the classes cannot be ordered with respect to one underlying dimension. The Latent Distance Model is one way of incorporating an hypothesized ordering on the latent classes into the model itself. There are other possibilities. *Located Class Models* assume the existence of m classes which are demarcated along an ordered continuum. One can suppose that within each class the local independence axiom is satisfied but that the trace functions have regressions which are monotonically related to the values at the end-points of these classes. They can then be supposed to be non-decreasing step functions. A linear relationship would serve as would any polynomial constrained to be non-decreasing over the latent continuum. These formulations of Located Class Models have a close relationship to the polynomial trace lines model but conditions for identifiability are harder to establish. Similar considerations apply to problems of ordering classes in the solution of a latent profile model. An examination of the latent profile solutions may generate plausible types of regression. Spacing in the pattern of mean values may suggest types of non-linear regressions for the manifest variables, or even a multidimensional latent space (Gibson, 1960). There is of course much indeterminacy in the method but after a reduction of the data by applying a latent class analysis, suitable forms may emerge.

Scaling models with trace lines defined differently from those considered have been developed for dichotomous manifest items in the context of test theory. A *normal ogive trace line* has been widely used (Lord, 1952; 1953). Here

$$\pi_j(y) = \frac{1}{\sqrt{2\pi}} \int_{-\infty}^{y_i} \exp\left[\frac{-(t-\mu)^2}{2\sigma^2}\right] dt.$$

The latent variable y is assumed to vary continuously over the real line and the form of the trace function ensures that it will always lie between zero and one. Models have also been developed using a normal density as the trace line itself.

No mention has been made of situations where the manifest variables are classifications with more than two responses. Here each item has several latent trace functions, one for each category, subject of course to the limitation that, conditional on a particular position in the latent space, their sum be one. A solution for one category for each item could be developed using the algebra for dichotomous items, by pooling the other categories together. If the model is valid consistent solutions will be obtainable. It appears then that an extension of the ordinary solution is all that is required, although the number of equations will become large leading to severe practical difficulties. These are discussed in Lazarsfeld and Henry (1968).

5.4. PARAMETER ESTIMATION, MODEL FITTING AND TESTING

So far it has been assumed that the joint population distribution of the manifest variables is known. The particular model under study assumed a true representation of the latent structure, and finding the parameters centres around the problems of specifying conditions and selecting equations so as to guarantee their identifiability. These are mathematical problems concerned with the consistency of the various distributions to enable the expression of the local independence relationship, with the specifications required to make the parameters of the model identifiable, and for identifiable structural parameters with ways of finding their values.

In a real world situation, however, one rarely expects the population latent structure model to hold exactly. One expects that it is an adequate approximation of the phenomenon under study, and even then, available data are usually sample based and consequently subject to sampling error so that they will not satisfy exactly the latent structure. The statistical problem here is one of discovering good estimation procedures. In addition, with sample data one will usually require procedures to test whether the hypothesized model fits the data adequately.

5.4.1 Estimation by the algebraic techniques

Several estimation procedures have been suggested for the parameters of the latent profile and class models, the most promising of which are those derived from *maximum likelihood* considerations. However, one initial point may be noted. If a latent class model is to be identifiable the parameters to be estimated can be expressed as continuous functions of the joint manifest probabilities. The sample manifest proportions are sufficient statistics for the population joint probabilities of the response patterns. This suggests replacing π_i, π_{ij}, π_{ijk} etc. in the basic accounting equations by their sample estimates $\hat{\pi}_i$, $\hat{\pi}_{ij}$, π_{ijk} etc., the observed joint proportions in the sample, and basing estimation on one of the algebraic solutions to the resulting equations. If items are chosen so as to make the various matrix inversions possible and conditions for identifiability are satisfied, this estimation procedure is at least consistent (i.e. gives estimates approaching the true values of the parameters as the sample size increases). This follows from the fact that the sample joint proportions are consistent for the population manifest joint probabilities. Functions of consistent estimators are also consistent for the same functions of those quantities being estimated, and the latent parameters through the accounting equations represent these functions. Any of the equation solving techniques introduced in section 5.3.1, Lazarsfeld and Anderson's basic solution, the modifications of Madansky using higher order proportions or the composite stratifier technique of Gibson can be used and consistent estimators ensue. It has also been shown (Anderson, 1968) that these estimators are asymptotically normally distributed and estimates of their variances and covariances can be derived, thus enabling in principle the construction of interval estimates and test procedures.

Only in the case of known manifest probabilities do these procedures give unique solutions, otherwise whilst computationally feasible they clearly neglect a good deal of relevant information. Further, they may give inadmissible solutions. Although they are consistent, due to the vagaries of data on moderate sized samples, particular estimates of latent probabilities may lie outside the range 0 to 1, and in certain circumstances these values may not even be real numbers. Also, different choices of the two sets of manifest variables and the stratifier variable in the basic solution, or different selections of equations in Madansky's or Gibson's extension may give many different, though possibly still admissible sets of estimates. Some further criteria for selection are obviously required. It has been pointed out by Anderson (1959) that the requirement in the basic solution for population data that Π_1 be non-singular indicates that it is preferable to choose items such that the corresponding matrix $\hat{\Pi}_1$ of sample proportions has a large determinant. This suggestion has been further supported by simulation studies. One approach then could be to find the value of this determinant for all selections of two sets of items, and select some number, say M, of these with the largest determinants. For each of the M combinations solve the equations using each possible stratifier, discarding at any stage solutions where any of the $\hat{\pi}_i^\alpha$ are not between zero and one. From the set of admissible solutions, a final estimate could be obtained by averaging. Alternatively, a measure of the goodness of fit of each solution to the model could be derived and the best fitting solution selected. For each manifest response pattern the observed proportions could be compared with those that the model would generate given each set of estimates and chi-square criterion calculated. That set giving the smallest chi-square value would be the final estimate. This is the basis of the algorithm developed by Lazarsfeld and named LASY (see Lazarsfeld and Henry, 1968).

5.4.2 Solutions by factoring methods

An alternative method of estimation based on the algebra of the accounting equations involves all first and second order data (Gibson, 1951; 1956), and adapts procedures normally associated with factor analysis. Here the problem of selecting among the lower order accounting equations does not arise. The first and second order accounting equations are cast in a form which enables the solutions of the equations to proceed by a consideration of how to factor the matrix $\mathbf{\Pi}$ into a form $\mathbf{F}'\mathbf{F}$. If such an $\mathbf{F}$ can be found its elements can be associated with certain latent parameters and solutions proceed.

The standard factor analytic procedures determine $\mathbf{F}$ only up to an orthogonal rotation. However, in latent structure models higher order accounting equations can be used to solve the problem of rotational indeterminacy. An algorithm for doing this has been devised by Green (1970). Mardberg (1971) and Isaacson (1972) have adopted Green's algorithm for the most widely used estimation procedure for the latent profile model known as LPA II. However, here the essential iterative factoring procedure is not incorporated, and this algorithm starts with the correlation and third order correlation matrix, being content to use autocorrelations or take some initial estimates of the unobservables as fixed, according to suggestions of Gibson (1959). A single iteration is then performed. This program does, however, allow the possibility of comparing different values of m up to maximum $p + 1$ permitted, postulating various measures of the fit of the latent classes when the individuals have been allocated to them by manifest information.

5.4.3 Estimation by maximum likelihood

The estimation procedures considered have been sample analogues of the various ways of choosing sets of equations for algebraic solutions of moment relations. They will, as has been pointed out, be consistent though will usually give different admissible solutions. However, there may be other consistent estimators which according to usual statistical criteria are more desirable, in the sense of having smaller asymptotic variance. One standard method of estimation is *maximum likelihood*, and although there are situations in which such estimators are not consistent, for the latent class model they are consistent. Furthermore, they have asymptotically a normal distribution and have in the generalized sense a smaller covariance matrix than any other normally distributed set of consistent estimators. The maximum likelihood estimators are then the 'best' asymptotical normally distributed set and are thus asymptotically efficient (McHugh, 1956). These BAN estimators for the latent class model are formed by maximizing the likelihood function for the latent parameters. The estimators are not expressible explicitly in terms of manifest data in a closed form, but for such situations iterative procedures are available using Taylor's approximation. An initial estimate is made and a series of successive approximations yields a solution. The procedure is often laborious and may require much computation but if feasible it is the most desirable. One needs an initial estimate of the parameters and since the values given by one of the algebraic solutions, or an average of a selection from a few are consistent, these seem to be most appropriate. An early attempt to approach the problem this way is formalized in the program BAN, written by Henry (1966).

A maximum likelihood approach to the estimation of parameters in other latent structure models is possible in principle but this would require assumptions about the forms of the distributions of the latent variable and the trace function. It may be postulated with appropriate assumptions that they would also satisfy the consistency and efficiency conditions holding for the latent class model. It is likely, however, that such procedures would give rise to great complexity and impracticability.

5.4.4 Estimation by partitioning

Another method of estimation explicitly uses the crucial assumption of local independence. Given a set of observations, if one knew the classes associated with them, then within a class the observations should show independence. This approach forms the basis of estimation by partitioning due to Mandansky (1959). For any arbitrary allocation of the n individuals to m classes in the latent class model, taking the dichotomous variables to be $x_i = 1, 0$ with within class probabilities π_i^α, $(1 - \pi_i^\alpha)$ the estimates would be

$$\hat{u}_\alpha = \frac{n(\alpha)}{n}$$

$$\hat{\pi}_i^\alpha = \frac{\text{Number of positive responses to } x_i \text{ in those allocated to } \alpha}{n(\alpha)}$$

where $n(\alpha)$ is the number allocated to class α; $n(\alpha)\, p(x_1, \ldots, x_n \mid \alpha)$ denotes the number of observations in the α'th class with response pattern $x_1, \ldots, x_p$ and $n(\alpha)$ $\hat{\pi}(x_1, \ldots, x_p \mid \alpha)$ would be that expected invoking local independence and the estimates $\hat{u}_\alpha$, $\hat{\pi}_i^\alpha$, $1 - \hat{\pi}_1^\alpha$. A chi-square goodness of fit measure would establish how well the proposed partition satisfied the local independence structure although Madansky suggests the easier to calculate measure

$$\Delta = \sum_\alpha \sum_{x_1, x_2 \ldots x_p} \frac{n^2(\alpha)(p(x_1, \ldots, x_p \mid \alpha) - \hat{\pi}(x_1, \ldots, x_p \mid \alpha))^2}{np(x_1, \ldots, x_p)}$$

Minimizing Δ gives an estimation procedure, but it would involve examining all possible partitions of the individuals, although an iterative procedure might be found. Although this procedure gives consistent estimates and is somewhat akin to minimum chi-square methods, it does not seem a practical one, except for small sample sizes and only a few latent classes. It could in principle be extended for the latent profile model although some grouping on the manifest scale would be necessary.

5.4.5 Problems of estimation

This section has presented some of the available techniques for estimation, and has hinted at some of their problems. Clearly two practical considerations are overwhelming; the computational feasibility of the techniques and problems of convergence when iterative procedures are used. These together with some idea of the precision of the various estimates have been the subject of some small scale theoretical and simulation studies. They leave one with the feeling that more work is really required at all levels to develop an entirely satisfactory estimation approach. The algebraic solutions developed by Lazarsfeld and Anderson are computationally simple but do not utilize much of the relevant information. There is some evidence from Monte Carlo studies and theoretical work that estimates are very imprecise for quite large samples, certainly up to 1,000, (Anderson, 1959). Green's extension of Gibson's factor method, whilst using much more of the available data and hence more information, and whilst avoiding the laborious problems of rotation, often gives a slow rate of convergence (Green, 1970). Again it appears that the estimates are very imprecise, i.e., have large variances. These results, using the L.P.A., model perhaps imply that in its most general formulation it is not very restrictive; i.e. different parameter structures often yield similar sample data. Again, maximum likelihood methods for the latent class model are inappropriate for sparse data. They require in their basic form a sample size sufficient to guarantee adequate data for all response patterns. For more than about four manifest variables the required sample size becomes quite large. Sampling experiments with the BAN iterative procedure (Lazarsfeld and Henry, 1968) have demonstrated there is often non-convergence when some item trace probabilities are close to zero or one, although convergence is often satisfactory when the latent class model holds exactly. Again large variances are noted and, as it seems this is a feature of all available methods, clearly caution must be exercised.

5.4.6 Evaluating goodness of fit

A supplementary problem arises in the specification, *a priori*, of the latent space. In the context of a latent categorization the number of classes, m, must be determined. Once the parameters have been estimated some method is required for evaluating the fit of the model, and particularly the chosen value of m, to the data. From earlier remarks about the rank of basic matrices in the algebraic solution, if population data were available, they would set a lower bound to m; similarly for the factor-orientated solution the rank of the correlation matrix sets this bound. An examination of adequate sample data should indicate the least number of latent classes appropriate. In practice different values of m may be chosen and the results compared. Higher order data not used in a solution, for instance, could be compared with that generated by the estimated parameters for each m and a best fitting value selected. From the full response pattern an obvious goodness of fit measure for the latent class model is

$$\chi^2 = \sum_{x_1, x_2, \ldots, x_p} \frac{[n(x_1, \ldots, x_p) - \hat{n}(x_1, \ldots, x_p)]^2}{\hat{n}(x_1, \ldots, x_p)}$$

where $n(x_1, \ldots, x_p)$ is the observed manifest number with response pattern x_1, $x_2, \ldots, x_p$, and $\hat{n}(x_i, \ldots, x_p)$ is the number expected to be generated by the structure using estimates of parameters. It would be useful if the distribution of this criterion, under the hypothesis that the specified number of latent classes was the true one, were known. This is not generally available but for maximum likelihood estimation this criterion is distributed as chi-square with $2^p - m\,(p + 1)$ degrees of freedom. In this case a test of $m = m_0$ could be made and if rejected then the hypothesized number set equal to $m_0 + 1$ in an effort to improve the fit. A sequence of fits could be made until the number of latent classes is adequate to fit the data.

In the LPA II implementation of the restricted form of Green's solution to the latent profile model, a series of *ad hoc* rules of thumb are suggested to determine the appropriate number of classes. Individuals are allocated to the classes derived in the solution by a distance criterion to be discussed shortly. Once this has been done one can form within class means m_i^α and variances s_{ii}^α for the allocated individuals.

$D_{st} = \Sigma_i(\hat{\mu}_i^\alpha - m_i^\alpha)^2$ is a measure of the closeness of the latent profile and manifest profile for class α, and $(1/m)\Sigma_\alpha D_\alpha$ is the total average square deviation measuring the fit between manifest and latent classes by their means. Other measures of fit have been proposed but this one seems quite appropriate. From some starting point m_0, latent class solutions are computed up to the maximum value possible with Green's solution, $p + 1$. That one is selected which gives the best fit according to this criterion (Mardberg, 1973; Isaacson 1972). Mardberg also suggested testing the series of within class correlation matrices against the identity matrix using a chi-square criterion and adding up across classes. However, the distribution of the criterion is not exactly known.

5.4.7 Allocating individuals to a position in the latent space

It may be the eventual aim of the analysis to allocate the individuals under study a latent scale value or class. Bayesian probability arguments could be used to do this, possibly subject to appropriate loss functions. The simplest method of doing this, if the trace functions and latent distribution have been completely specified, is given by (2). For the latent class model, for instance,

$$P(\alpha \mid X_1, X_2, \ldots, X_p) = \frac{p(X_1, X_2, \ldots, X_p \mid \alpha)u_\alpha}{p(X_1, X_2, \ldots, X_p)}$$

where $X_1, X_2, \ldots, X_p$ with $X_i = 0, 1$ is the individual's observed response pattern. A decision rule could be to maximize the above and allocate an individual to the most likely class. This could be extended to non-dichotomous X_i. In situations where the distributions are only partly specified, other criteria for assignment must be developed. Thus although an assignment using equation (2) could be used for the trace line models when the probability density over the latent space is specified, this is not the case when only a few moments have been specified. In its least restrictive form, the latent profile model leaves trace functions partly specified although if the additional assumptions of within class normality could be made an allocation follows from knowledge of means and variances. Without such assumptions a criterion based on distance measures seem justifiable. Suppose an analysis has proceeded in terms of standardized variables $Z_i = (x_i - \mu_i)/\sqrt{\sigma_{ii}}$ so that μ_i^α and σ_{ii}^α are also in standard units. The distance measure of an individual from a latent class could be any of those used in the context of cluster analysis although for the latent profile model it has been suggested that

$$d_{j\alpha} = \sqrt{\sum_{i=1}^{p} (X_{ij} - \mu_i^\alpha)^2}$$

is a suitable measure where X_{ij} is the value of variable x_i for individual j. A person is assigned to that class α for which $d_{j\alpha}$ is minimum. However, other measures are possible, and perhaps a more appropriate one (certainly for normal trace distributions) utilizing differential class variances is:

$$d_{j\alpha}^{*} = \sqrt{\Sigma\left(\frac{X_{ij} - \mu_i^{\alpha}}{\sigma_{ii}^{\alpha}}\right)^2},$$

although this should make little difference if within class variances are similar. Assignment of respondents can thus be viewed as a way of clustering individuals according to a predefined classification, assigning a scale value or ordering along a continuum.

5.5 EXAMPLES

In this section the usefulness of latent structure techniques is illustrated by reference to a few published applications.

Lazarsfeld and Henry (1968) present many of the applications of various formulations of the model based on their own and other research. An interesting example which demonstrates the possible uses of latent structure approaches in the area of medical diagnosis attempts to classify groups on the basis of differential mortality from manifest data on a variety of observable variables known to be related to mortality, (Miller, Sabagh and Dingman, 1962). The data were collected on 3709 patients admitted between 1949 and 1956 to a hospital for the mentally retarded. Some of the variables were continuous but dichotomized suitably they were: sex (male/female), age (under and over 10), ethnic status (caucasian/non-caucasian), I.Q. (under and over 40), diagnosis (somatic/non-somatic). From previous substantive findings a two-class solution was proposed and Green's estimation method was applied with the results in Table 5.1.

A comparison of the observed second and third order joint manifest proportions was made with the expected proportions generated by the estimated parameters. Rough chi-square measures were adopted and it was found that the fit was very good. Respondents were also allocated to the most likely class on the basis of their set of observations and the observed proportions for each manifest characteristic, and

Table 5.1. Latent probabilities for the Miller, Sabagh and Dingman data (Reproduced with permission from C. R. Miller, H. Sabagh and H. F. Dingman, *Journal Amer. Statist. Ass.*, 57 430–438 (1962))

	Class 1	Class 2
Female	0.40	0.45
Under 10	0.15	0.72
Caucasian	0.65	0.93
Under 40 IQ	0.25	0.92
Somatic diagnosis	0.28	0.89
Proportion of patients in each class	0.60	0.40

Table 5.2. Deaths per 1,000 person-years by duration of hospitalization and by latent class, Pacific State Hospital, 1944–1956[a] (Reproduced with permission from C. R. Miller, H. Sabagh and H. F. Dingman, *Journal Amer. Statist. Ass.*, **57**, 430–438 (1962))

	Latent class	
Duration of hospitalization	1	2
0–3 months	26(13)	288(113)
3–6 months	11(4)	146(54)
6–9 months	6(3)	70(25)
9–12 months	15(7)	51(18)
1–2 years	3(6)	29(39)
2–3 years	6(10)	22(29)
3–4 years	1(2)	21(27)
4–5 years	3(5)	16(19)

[a]The number of deaths on which the rates are based are given in parenthesis.

the proportion in each class was compared to those of Table 5.1, and the fit was again close. It was concluded that the two classes represented mildly retarded, class 1, and severely retarded, class 2. The first class was mostly under 10, white, under 40 I.Q. and with somatic diagnosis. Sex as might be expected did not differentiate between the classes and might possibly be dropped from the analysis. The initial aim of this study was to develop some compound classification of respondents on the variables in order to study how they were related to other factors affecting mortality rates. Thus one has two homogeneous classes formed as an aid to further analysis. In fact the effect of duration of hospitalization on mortality rates was of interest and also whether this interacted with other personal characteristics. Using only two classes to represent compound classifications instead of all configurations ensured that there was a sufficient number of cases to allow computation or mortality rates for small hospitalization periods. The results show, as is expected, generally higher rates for class 2 than class 1, but this differential decreases with increasing length of hospitalization. Rates generally decrease for both classes as hospitalization period increases. Table 5.2 shows these results.

This application illustrates a simple but potentially valuable use of latent structure models where it is desired to evaluate the effects of several characteristics taken together. Here the actual number of deaths was too small to permit extensive cross tabulations, and this problem of 'sparsity' of data is often a feature of survey work when data is qualitative. The latent structure approach permits the formulation of relatively homogeneous classifications.

Latent Profile Models have also been used extensively as an approach to the common problem of classifying objects into homogeneous clusters. Examples in the context of industrial psychology where data from job analysis surveys are used to select and classify personnel are given by Mardberg (1973). Here the search is for general requirement profiles for various types of job and latent profile analysis is

Table 5.3. Means and standard deviations for eight rating variables. $n = 37$. (Reproduced with permission from B. Mardberg, *Clustering Jobs to General Requirement Profiles*, Swedish Council for Personnel Administration, 1972)

Variable	Mean	Standard deviation
1. Experience	2.05	0.62
2. General ability	2.03	0.60
3. Spatial imagination	2.03	0.64
4. Spatial coordinating ability	1.97	0.55
5. Attentiveness	2.24	0.44
6. Perceptual vigilance	1.89	0.66
7. Motor coordination	2.00	0.47
8. Reaction ability	1.97	0.44

used in the initial phase of the research, and as part of the development of a larger model for selection, classification and placement of individuals. In this work the motivation for the use of latent structure models was the implicit recognition of possible non-linear relationships between manifest and latent variables, and their resulting greater generality. Here latent profile analysis is used as a classification procedure in preference to other methods of clustering based on inter-object distances or similarities, such as correlations between objects (individuals) over variables. Some of these other methods such as Q-mode factor analysis and single-linkage clustering are discussed elsewhere in this volume (Chapters 3 and 4). Apart from the advantage implied by the possibility of nonlinearities, a latent profile approach uses all the manifest information, instead of simple similarity indices, and enables analysis to be related to a much larger number of study objects.

In one of Mardberg's applications 37 different foundry jobs at an ironworks in Sweden formed the study units. Each job was studied for a period of two weeks and ratings for each were made on a three-point scale for each of eight psychological requirement variables (Mardberg, 1972). Thus, although the manifest data were discrete with low requirements, normal requirements, and high requirements scored 1, 2, 3 respectively, no distributional assumptions need be made but

Table 5.4. Intercorrelations between rating variables (product moment) $n = 37$. (Reproduced with permission from B. Mardberg, *Clustering Jobs to General Requirement Profiles*, Swedish Council for Personnel Administration, 1972)

Variable	2	3	4	5	6	7	8
1. Experience	0.96	0.69	0.65	0.26	0.35	0.38	0.41
2. General ability		0.64	0.59	0.19	0.36	0.39	0.42
3. Spatial imagination			0.70	0.37	0.53	0.55	0.59
4. Spatial coordinating ability				0.49	0.60	0.75	0.57
5. Attentiveness					0.58	0.41	0.33
6. Perceptual vigilance						0.54	0.57
7. Motor coordination							0.54
8. Reaction ability							

Table 5.5. Latent profiles. (Reproduced with permission from B. Mardberg, *Clustering Jobs to General Requirement Profiles*, Swedish Council for Personnel Administration, 1972)

Variable	Profile 1	2	3
1. Experience	−1.19	0.01	1.01
2. General ability	−1.16	−0.05	1.02
3. Spatial imagination	−1.14	0.73	0.60
4. Spatial coordinating ability	−1.14	1.09	0.41
5. Attentiveness	−0.58	1.45	−0.26
6. Perceptual vigilance	−0.83	1.47	−0.05
7. Motor coordination	−0.89	1.25	0.12
8. Reaction ability	−0.89	1.06	0.22
Estimated proportion of jobs	0.361	0.218	0.421
Estimated number of jobs	13.4	8.1	15.6

underlying continuity could be assumed. It is possible that a more appropriate approach might have been to regard the manifest variables as trichotomies and to apply the extension of the latent class model.

The eight variables, which are defined more fully by Mardberg are: experience, general ability, spatial imagination, spatial coordination ability, attentiveness, perceptual vigilance, motor coordination and reaction ability. A latent profile analysis using the LPA restricted version of Green's method of the standardized

Table 5.6. Observed profiles. Means and standard deviation of raw scores. (Means in the first row). (Adapted with permission from B. Mardberg, *Clustering Jobs to General Requirement Profiles*, Swedish Council for Personnel Administration, 1972)

Variable	Profile 1	2	3
1. Experience	1.33	2.14	2.33
	0.50	0.38	0.48
2. General ability	1.33	2.00	2.33
	0.50	0.00	0.48
3. Spatial imagination	1.22	2.57	2.19
	0.44	0.53	0.40
4. Spatial coordinating ability	1.33	2.57	2.05
	0.50	0.53	0.22
5. Attentiveness	2.00	3.00	2.10
	0.00	0.00	0.00
6. Perceptual vigilance	1.22	2.71	1.90
	0.44	0.49	0.44
7. Motor coordination	1.67	2.43	2.00
	0.50	0.53	0.32
8. Reaction ability	0.53	0.49	0.22
Number of jobs	9	7	21

Table 5.7. Correlations between contact characteristics: meetings and telephone calls combined

	Media	Length	Arrangement	Frequency	Feedback	Range	Trading	Participants
1. Media	1.00							
2. Length	0.70	1.00						
3. Arrangement	0.65	0.60	1.00					
4. Frequency	−0.17	−0.20	−0.13	1.00				
5. Feedback	0.31	0.29	0.29	−0.07	1.00			
6. Range	0.27	0.33	0.21	0.03	0.23	1.00		
7. Trading	−0.04	0.05	−0.02	−0.09	−0.01	0.01	1.00	
8. Participants	0.52	0.66	0.50	−0.15	0.20	0.22	0.04	1.00

correlation and third order correlation matrices was adopted. On the basis of the *ad hoc* measures of fit a three-class solution was proposed, and this was considered a good one after checking other measures. The manifest summary data are shown in Table 5.3, with correlation matrix in Table 5.4.

The latent profile solution is given in Table 5.5 in terms of standard scores. Once an assignment of jobs to classes has been made, the manifest profiles may be computed, in addition to observed within class variances. These are shown in raw score form in Table 5.6. Most of the within class deviations are small and from the *ad hoc* measures the fit of the model seems good. The first class is characterised by relatively low requirements throughout, except for attentiveness, and the jobs classified are unskilled. The second class has high requirements especially on attentiveness and perceptual vigilance. The third class has high requirements with respect to experience, general ability, and spatial imagination, otherwise they are moderate, and jobs classified are skilled, require long experience, and good abilities to understand the production process. Mardberg concluded that the results indicated the feasibility of forming general requirement profiles from judgements of requirements for single jobs. He also tried a clustering scheme based on principal component scores and similar results emerged, suggesting perhaps very small non-linearities in the underlying relationships. This example demonstrates the usefulness of LPA as an approach to non-linear based reduction of data space.

Latent profile analysis has also been used to construct a detailed location picture of office employment in Central London. One part of the study collected detailed information on contact with other people from a sample of senior personnel in the study firms. Full details of the study methodology can be found in the report (Goddard, 1971). Latent profile analysis was used to develop a typology of contracts based on their recorded characteristics. Eight characteristics common to both telephone and meeting contacts were used: length, prearrangement (5 point scale), frequency (5 point scale), purpose (one way/two way interaction), range of subjects discussed (3 point scale), concern with purchase or sale of goods and services (3 point scale), number of participants involved, media used (1 = telephone, 2 = meetings). Data on 6680 contacts were used, and an initial two class solution was tried. Relevant information is presented in Tables 5.7 and 5.8.

Table 5.8. Two profiles: meetings and telephone calls combined. (Class means in standardized units)

	Class I	Class II
1. Media	1.90	−0.33
2. Length	1.96	−0.34
3. Arrangement	1.61	−0.28
4. Frequency	−0.57	0.10
5. Feedback	0.77	−0.13
6. Range	0.81	−0.14
7. Trading	0.11	−0.02
8. Participants	1.53	−0.27
Number of contacts in each Class	994	5686

Class 1 contacts were generally meetings, of above average length, arrangement, and number of people involved, but below average frequency, possibly involving unfamiliar participants. Class II were identified as generally telephone contacts, of below average length, arrangement and number of people. Also they are marginally below average on feedback and range, while having above average frequency. In accordance with theory the classes were labelled orientation contacts and programmed contacts. The manifest profiles for allocated contacts are shown in Table 5.9 and by an ingenious pictorial device in Figure 5.1. No mention is made in the report of testing the adequacy of the model but at least a three-class model was also tried. This apparently was justified by an appeal to theory which suggests an intermediate type of contact, planning contact (Thorngen, 1967). The new class is characterized by contacts that are generally very short, frequent, involve only two

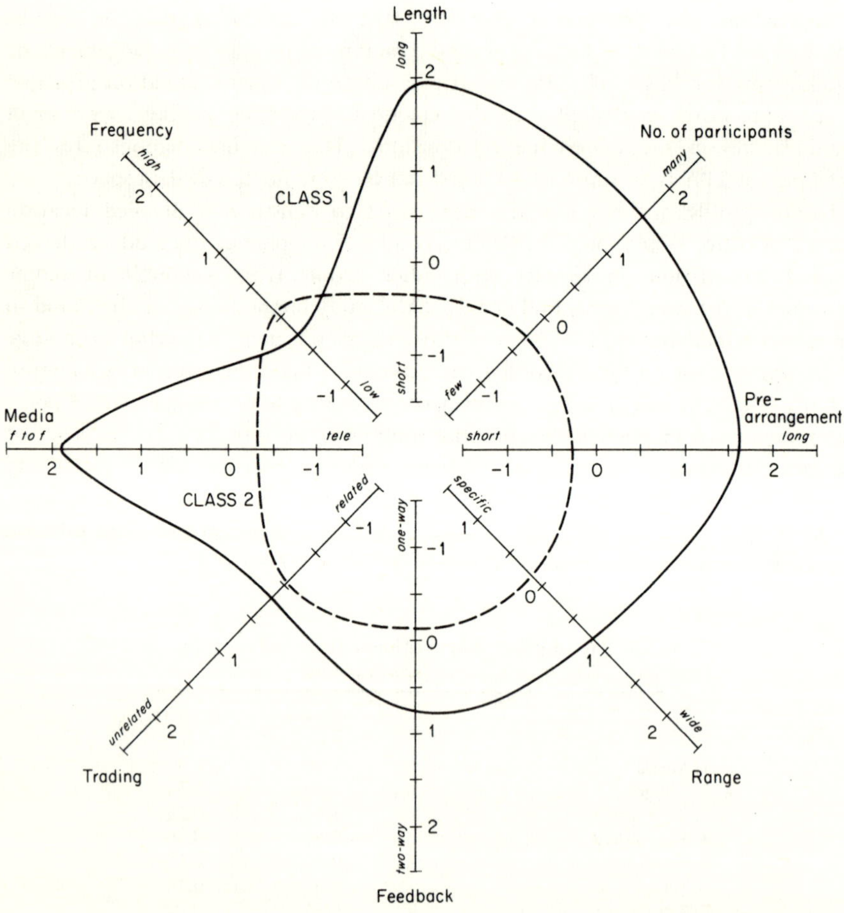

Figure 5.1. Classification of face-to-face meetings and telephone contacts using Latent Profile Analysis (2 profiles).

Table 5.9. Means and standard deviations in raw score units for two profiles: meetings and telephone calls

		Class I	Class II
1. Media	Mean	2.0	1.1
	S.D.	0.09	0.27
2. Length	Mean	3.4	1.2
	S.D.	1.15	0.38
3. Arrangement	Mean	3.9	1.4
	S.D.	1.11	0.94
4. Frequency	Mean	2.2	3.1
	S.D.	1.08	1.33
5. Feedback	Mean	1.7	1.3
	S.D.	0.45	0.46
6. Range	Mean	1.7	1.2
	S.D.	0.74	0.41
7. Trading	Mean	2.1	2.0
	S.D.	0.84	0.88
8. Participants	Mean	1.8	1.0
	S.D.	0.92	0.05

participants, but with lengthy prearrangements. These require a fair amount of feedback and are directly concerned with sales and purchases. A few contacts have been reallocated from the original two classes but the basic characteristics of these remain substantially as before. Again no attempt was made to assess the fit of the model but the results do seem to verify the theoretical notions of types of contact. This typology was also used to examine differences between types of department, industrial sectors, and status of person in the type of contact. Thus at least in this application the latent profile was used empirically to verify a theoretical typology, and at the same time used the empirically constructed typology in further analyses relating to other aspects of the survey data. This was done by examining differences in the proportions of contacts allocated to each class in categories of other characteristics. This is a similar use of constructed classes to that in the investigation of differential mortality discussed earlier.

Apart from a direct use of latent structure models as a scaling technique of which many examples are reported elsewhere (Lazarsfeld and Henry, 1968), a recent review has discussed the usefulness of the local independence model to panel data (Wiggins, 1973). Here the manifest data are usually on relationships between responses to one variable only taken at several points of time. It is possible here to assume also that individuals change their latent position over time.

The approach here was mainly concerned with latent dichotomies and manifest dichotomies. Various types of model have been applied to panel studies of voting intention and attitude changes. As a simple example, Wiggins analysed the two wave panel data on voting intention given in Table 5.10 from a study of the election campaign in Elmira, New York, using data from interviews in August and October and dichotomizing responses into 'Republican' and 'Not Republican'. There are assumed to be two latent classes arbitrarily called 'True Republican' and 'Not True

Table 5.10. Vote intention, August and October, Elmira Panel (Reproduced with permission from L. Wiggins, *Latent Probabilities for Attitude and Behaviour Processes*, Elsevier, Amsterdam, 1973)

		October		
		Republican	Not Republican	Total
August	Republican	353	20	373
	Not Republican	22	167	189
	Total:	375	187	562

Republican'. People in each class have a latent probability of declaring their true position which is assumed to be the same for both classes. The latent classes are assumed to remain the same over time but the latent probability can change. Using algebraic approaches, Wiggins found the latent parameters of Table 5.11. The latent probability that a respondent in a given latent class will declare his true class is quite high on both occasions. Other complex approaches to multiwave data also embodying assumptions about the latent parameter changes are discussed in depth by Wiggins. Other recent work of interest has been latent structure modelling of item sorting in similarity assessments (Craig, 1973). Harper (1973) investigates the possibility of relaxing somewhat the local independence axiom to apply to situations where pairwise local dependency is thought to hold.

Table 5.11. Latent structure solutions (Reproduced with permission from L. Wiggins, *Latent Probabilities for Attitude and Behaviour Processes*, Elsevier, Amsterdam, 1973)

Time	Latent probability of true declaration	Proportion in Latent Class 1 (Republican)	Proportion in Latent Class 2 (Not Republican)
1	0.956	0.686	0.314
2	0.966	0.686	0.314

REFERENCES

Anderson, T. W. (1954), 'On estimation of parameters in latent structure analysis', *Psychometrika*, **19**, 1–10.

Anderson, T. W. (1959), 'Some scaling methods and estimation procedures in the latent class models', in Grenander, U., ed., *Probability and Statistics*, New York: Wiley.

Anderson, T. W. (1968), 'Large sample distribution theory for estimates of the parameters of a latent class model', Appendix B in Lazarsfeld, P. F., and Henry, N. W., *Latent Structure Analysis*, Boston: Houghton Miflin.

Gibson, W. A. (1951), *Applications of the Mathematics of Multiple Factor Analysis to the Problems of Latent Structure Analysis*, unpublished Ph.D. dissertation, University of Chicago.

Gibson, W. A. (1955), 'An extension of Anderson's solution for the latent structure equations', *Psychometrika*, **20**, 69–73.

Gibson, W. A. (1956), 'Proportional profiles and latent structure', *Psychometrika,* **21**, 135–144.

Gibson, W. A. (1959), 'Three multivariate models, factor analysis, latent structure analysis, and latent profile analysis', *Psychometrika,* **24**, 229–252.

Gibson, W. A. (1960), 'Non-linear factors in two dimensions', *Psychometrika*, **25**, 381–392.

Goddard, J. B. (1971), *Office Communications Patterns in Central London*, Vol. II of *Office Linkages in Central London*, South East Economic Planning Council.

Green, B. F. (1951), 'A general solution for the latent class model of latent structure analysis', *Psychometrika,* **16**, 151–166.

Green, B. F. (1952), 'Latent structure analysis and its relation to factor analysis', *J. Am. Statist. Ass.*., **17**, 71–76.

Green, B. F. (1970), 'Parameter estimation in latent class analysis', *mimeo*, Department of Psychology, Baltimore: John Hopkins University.

Harper, D. H. (1973), 'Relaxing the axiom of local independence', *Psychometrika*, **38**, 273–286.

Henry, N. W. (1966), *BAN, A Fortran Program* (available from the Computing Laboratory, Cambridge as NEWBAN).

Hays, D. G. and Borgatta, E. F. (1954), 'An empirical comparison of restricted and general latent distance analysis', *Psychometrika*, **19**, 271–279.

Isaacson, A. (1972), 'Latent profile analysis; a brief description', *Technical Report No. 54, Royal Institute of Technology,* Stockholm.

Koopmans, T. C. (1949), 'Identification problems in economic model construction', *Econometrica*, **17**, 125–138.

Koopmans, T. C. and Reiersøl, O. (1950), 'The identification of structural characteristics', *Annals of Mathematical Statistics*, **21**, 165–181.

Lawley, D. N. and Maxwell, A. E. (1971), *Factor Analysis as a Statistical Method* (2nd Ed.), London: Butterworth.

Lazarsfeld, P. F. and Henry, N. W. (1965), 'The application of latent structure analysis to quantitative ecological data', *Mathematical Explanations in Behavioral Science*, Massarik, F., and Ratoosh, P. (eds.), Homewood, Illinois: Dorsey Press.

Lazarsfeld, P. F. and Henry, N. W. (1968), *Latent Structure Analysis,* New York: Houghton Mifflin.

Lord, F. M. (1952), 'A theory of test scores', *Psychometric Monogragbs*, No. 7.

Lord, F. M. (1953), 'The relation of test scores to the trait underlying the test', *Educ. Psychol. Measure.*, **13**, 517–549.

Madansky, A. (1959), 'Partitioning methods of latent class analysis', *Report P 1644*, Rand Corporation,.

Madansky, A. (1960), 'Determinental methods in latent class analysis', *Psychometrika*, **25**, 183–198.

Mardberg, B. (1963), 'L.P.A., a preliminary program for the solution of latent profile analysis, according to Green's method', *mimeo*, University of Stockholm.

Mardberg, B. (1972), *Clustering Jobs to General Requirement Profiles,* Stockholm: Swedish Council for Personnel Adminstration.

Mardberg, B. (1973), 'A model for selection and classification in industrial psychology', *Supplement 19*, Psychological Laboratories, University of Stockholm.

Marshcak, J. (1942), 'Economic interdependence and statistical analysis', *Studies in Mathematical Economics and Econometrics*, 135–150. University of Chicago Press.

McDonald, R. P. (1962), 'A general approach to non-linear factor analysis', *Psychometrika*, **27**, 397–415.

McHugh, R. B. (1956), 'Efficient estimation and local identification in latent class analysis', *Psychometrika*, **21**, 331–347.

Miller, C. R., Sabagh, H. and Dingman, H. F. (1962), 'Latent class analysis and differential mortality', *J. Am. Statist. Ass.*, **57**, 430–438.

Somers, R. H. (1961), *Latent Content Model of Latent Structure*, Ph.D. dissertation, Columbia University.

Thorngen, B. (1967), 'Regional economic interaction and flows of formation', Regional Science Association, *mimeo*.

Wiggins, L. (1973), *Latent Probabilities for Attitude and Behaviour Processes* Amsterdam: Elsevier.

Chapter 6

Multidimensional Scaling

A. P. M. Coxon and C. L. Jones

6.1 INTRODUCTION

Our purpose in this Chapter is threefold:

(i) to show that a geometric representation provides a natural and useful framework for analysing social science data in general, and survey data in particular
(ii) to provide an overview of multidimensional scaling (MDS) models, outlining the ways in which the original approach has been generalized to deal with a wide variety of models and types of data
(iii) to illustrate the applicability of such models to problems of analysis that routinely confront survey researchers by analysing the responses of a large heterogeneous sample to a set of 21 questionnaire items dealing with 'attitudes to work', by means of the Carroll–Chang INDSCAL model.

6.1.1 Survey Data

Scaling methods are best thought of as part of the survey analyst's armoury of descriptive techniques. Multidimensional scaling (MDS) can be regarded as the extension of methods already familiar in survey research, such as Guttman's scalogram analysis, Coombs, unfolding principle and factor analysis. The fundamental problems which arise in scaling are those of model building and testing, but the models dealt with here differ from the currently fashionable 'causal models' of variable-centred analysis. One word of warning should be emphasized before we start however. The sampling properties of almost all MDS models and procedures are little known, although a start has been made on the problem, (Barlow *et al.* (1972)). MDS procedures should therefore be used with the caution one reserves for any very powerful but unfamiliar procedure.

Having carried out a survey of people, or of dwellings, or of business firms or other units, the researcher will often wish to describe the characteristics of the 'average' or 'typical' respondent, household, firm and so forth. He will also wish to give his readers some feeling for the range of variation around this average. Where he is concerned with only one variable, such as the size of a family, or the value of a firm's liquid assets, it is possible to describe the range and type of variation by means of the usual statistical indices of central tendency and dispersion, together with graphical methods for displaying the shapes of frequency distributions. But the distributions of variables are described one at a time, and separately from one another, and since social scientists are concerned with patterned relationships between variables, they do not always find this sort of procedure very useful. The main problem is often one of making informative statements about the ways in which the subjects of a survey are similar to one another (cluster together) and conversely about the ways in which they differ, and this over sets of variables, not just over one variable at a time. Economists have discussed this as the 'aggregation problem' and psychologists refer to it as 'individual differences'. In each case the problem is the same: if subjects systematically differ in terms of the processes underlying the data, then it will not be possible to recover this fact after they have been summarily aggregated. In non-experimental and exploratory studies, where the general form rather than the exact detail of the underlying processes is the primary concern, we ought to pay special attention to the extent to which, and the ways in which, individual differences exist and are likely to affect our inferences. Not all multidimensional scaling models are concerned to represent individual differences, but we believe that those which can do so will be increasingly used by survey researchers since they provide an 'aggregating' framework in terms of which patterns which exist in the data of individuals (or firms, or whatever) are still represented in summary descriptions. Individual differences scaling approaches should make these descriptions more sensitive, and at the same time, more precise, distinguishing between the attributes that all the subjects of the survey have in common from those in which they differ in a patterned way, and distinguishing these in turn from random variation.

6.1.2 Multidimensional Scaling

Multidimensional scaling (MDS) has a shorter history than factor analysis, but it was equally restrictive in its original form. In the case of 'classic MDS' (expounded in detail in Torgerson (1958)) the basic data usually consisted of judgments of the psychological similarity or dissimilarity between each pair of a set of stimuli, and were obtained from a number of individuals. Torgerson describes an illustrative study of colour perception, where these judgments were obtained by what is called the 'complete method of triads'. The 84 possible triads which can be formed from 9 stimuli were set up, and each subject was asked to say which of the remaining two colours was more similar to the third one, for each triad. This illustrates one of the major disadvantages of 'classic MDS' – the vast amout of data systematically collected from each subject. In Torgerson's study of colour perception, complete data were collected from each of 38 subjects and all these data were pooled in such a way that for each colour there was an empirical estimate of the proportion of times its was judged as more similar to colour *i* than to colour *j*. Assuming that it is useful to try to represent such data by a distance model, these proportions are then interpreted as being in some way related to distances in a one, two, three or higher, *r*-dimensional, space. The most obvious analogy here is that of an ordinary road map, where Blackpool is in fact nearer in mileage distances to London than it is to Edinburgh, and we can find this out by comparing the distance between Blackpool and London with the difference between Blackpool and Edinburgh, i.e. by looking at a difference between distances. 'Classic MDS' then estimates the absolute distances between all pairs of stimuli from the original data by least squares methods which need not concern us. Returning to the analogy of the road maps, the outcome was equivalent to that useful table which shows the mileage distance between any pair of towns. 'Classic MDS' proceeds from this point by converting the matrix of absolute distances into a matrix of scalar products which are then treated as covariances and subjected to factor analysis (with all the usual problems about dimensionality and so forth), in order to obtain the projections of the stimuli upon as many orthogonal axes as the user wishes. In the road map analogy, the data on distances between all pairs of towns are converted without loss of any information into a 'solution' consisting of the position of each town on each of two reference axes. In Torgerson's study, the nine colours used as stimuli were all of the same red hue, but varied in terms of brightness and saturation. The subjects' judgments of similarity reflected this, since it was possible to rotate the two factor solution so that the projections of the stimuli on the two factors reflected the (already known) values of the stimuli on brightness and saturation respectively.

It should be evident that the methods of 'classic MDS' were exceedingly wearing upon the patience of experimental subjects, and also involved vast amounts of tedious computation. Developments in this area would have remained of academic interest to survey analysts but for two things. First, a methodological shift of focus and a technical breakthrough occurred in the early 1960s which made MDS procedures practical for large sets of data at virtually any level of measurement, and secondly, the basic model was quickly extended to deal with types of data commonly encountered by survey analysts. The methodological viewpoint associated in the main with Coombs is developed in detail in *A Theory of Data* (1964) and his advocacy of non-metric approaches to measurement might be summarized as follows:

(i) assumptions which are made about the 'level of measurement' of one's data and those contained in the scaling models used to analyse them are not merely convenient fictions, but commit one to substantive hypotheses about human behaviour.

(ii) It is better to err on the side of conservatism in attributing metric properties to social science data, and use weaker measurement structures to represent them. A cautious approach is also desirable when making assumptions about the population distributions of multivariate data.

(iii) Because most social science data are elicited in non-experimental settings, and often refer to 'diversified' or non-homogeneous populations, it is well to be especially sensitive to individual differences which may be crucial to the interpretation of one's data.

The technical breakthrough occurred when Shepard (1962, 1966) went on to demonstrate that merely non-metric constraints, (i.e. ordinal information in the data) when imposed in sufficient number, put very tight limits indeed upon the possible solutions to the data. (By 'solutions' in this context, one usually means the projections of the stimulus points on a number of reference axes: e.g. the latitudes and longitudes of the stimulus points). So tight are the limits on the solution imposed by having a large number of ordinal constraints that it is possible to identify the best *metric* solution (the projections of stimulus points upon the reference axes being defined at an equal-interval level of measurement) for merely ordinal data.

6.2 Non-metric multidimensional scaling

The greatest advantage of non-metric MDS is that the same basic procedure can be readily extended to very different types of data and to different models (other than the usual distance model). The popularity of this extension can be gauged from the fact that applications soon appeared in disciplines as diverse as electronics (Kruskal, 1966) and history (Hodson *et al.* 1971), although social science applications continue to predominate.

A further advantage of non-metric MDS is that, unlike conventional multivariate statistical models, non-metric MDS solutions are 'order invariant' (Sibson, 1972) – i.e. any measures having the same *ordering* of data will give rise to the same solution, and the procedures in no way depend upon the particular measure of similarity used. To the extent that data can be construed as providing information on the relative proximity between points (which is precisely Coombs' way of interpreting most types of data) to this extent it may reasonably be represented by the distance model. Nor need the data be composite measures, such as the covariances and correlations of factor analysis; measures such as contingency coefficients between attributes, co-occurrences of words in a text, and confusions between sounds in a listening task are equally legitimate. The numerous ways of making up indices of resemblance between individuals or between variables have been documented by Cormack (1971; p. 324 on), by Jardine and Sibson (1971; part 1), and by Sokal and Sneath (1963; Chapter 6). Moreover, several MDS models have been evolved which keep individual data intact, and this represents a significant advance over factor analysis.

For example, if an individual rates or ranks a certain number of stimuli in terms of preference (or cognate terms) his data are kept together in the Carroll and Chang (1964) MDPREF model (see below for an example). In a similar way, if an individual makes judgments about the degree of similarity between each pair of a number of stimuli, his data are kept together in the Carroll and Chang (1970) INSCAL model (see below for an example). (By 'kept together' here, we mean that the scaling models contain parameters representing each individual, as well as for each stimulus point).

How does non-metric MDS work? Shepard (1966) begins to explain the rationale by considering the fact that the numerical representation of a set of objects is relatively indeterminate when only the ordering of the objects are known. This indeterminacy stems from the fact that the position of the points representing the objects in the solution space can be moved very extensively (i.e. can take on a large range of numerical values) and still satisfy the purely ordinal constraints represented by the data. (Whether or not these numerical values turn ot to be unidimensional remains to be decided). However, once the location of points in the solution must in addition satisfy constraints which refer to the order of interpoint distances, then the range of possible numerical values is greatly reduced:

> 'If non-metric constraints are imposed in sufficient number they begin to act like metric constraints . . . As the points are forced to satisfy more and more inequalities on the interpoint distances the spacing tightens up until any but very small perturbations of points will usually violate one or more of the inequalities' (p. 288)

The notion that *order* relations on distances impose severe constraints on the uniqueness of numerical representation is now commonplace, but its convincing demonstration awaited the development of an iterative algorithm designed to produce a solution which satisfied the set of constraints obtained from the data.

The basic rationale for the non-metric MDS algorithm was given by J. B. Kruskal (1964) and this has formed the basis for almost all subsequent work in this area. The goal of any non-metric procedure (at least for a distance model) is to find a set of points in a space of minimum dimensionality such that the dissimilarities data are a *monotone (ordered) function* of the distances in this space, i.e. that when the dissimilarity between stimulus i and j is less than the dissimilarity between k and l, then the distance between i and j should be at least as small as that between k and l: i.e.

$$\text{whenever} \quad \delta_{ij} < \delta_{kl} \quad \text{then} \quad d_{ij} \leqslant d_{kl}$$

(the so-called Weak Monotonicity criterion), where δ_{ij} is the dissimilarity between point i and point j, and d_{ij} is the corresponding distance. A configuration of points in an r-dimensional space which satisfies this criterion for a given set of data is termed an r-dimensional solution.

A straightforward application of non-metric MDS, in a sociological context, has been reported by Macdonald, (1972), as part of an investigation into the dimensions underlying the 1951 British census classification of occupations into thirteen 'socio-economic groups'. Since he was interested in mobility from father's occupation to son's occupation, Macdonald chose to base his index of dissimilarity, (between each pair of the thirteen socio-economic groups), upon the cross-tabulation of each son's socio-economic group against that of his father. A tabulation of this kind from the 1951 Census had been published by Benjamin (1958), and part of it is shown here as Table 6.1. The first row of the table shows the proportions of sons of agricultural workers who ended up in the different socio-economic groups, 3.4 per cent entering becoming farmers (SEG 1), 40.0 per cent following in their fathers' footsteps, 0.6 per cent entering higher administrative occupations (SEG 3), and so forth. Each row of the table contains information on the proportions of sons entering each socio-economic group, given that their fathers were employed in a particular socio-economic group.

The measure of dissimilarity between socio-economic groups as used for classifying occupations of fathers is simple and even crude. For any pair of rows of

Table 6.1. Percentage of sons entering 13 occupational groups, for three groups of fathers (Reprinted with permission from B. Benjamin, *Population Studies*, **11**, 262–268 (1958))

Socio-economic group of father	Socio-economic group of son													Total (100%)
	1	2	3	4	5	6	7	8	9	10	11	12	13	
2. Agricultural workers	3.4	40.0	0.6	2.2	2.8	0.0	1.7	2.2	1.7	26.4	5.1	13.5	0.5	178
5. Shopkeepers	2.5	2.5	5.7	14.7	27.9	7.4	9.0	0.8	1.6	17.2	3.3	6.6	0.8	122
10. Skilled workers	0.3	2.8	2.4	7.2	4.1	5.7	2.5	1.6	3.8	47.0	12.0	9.9	0.6	996

Table 6.1, one computes the absolute differences between corresponding entries. The sum of these absolute differences is an index of dissimilarity between whatever pair of 'father's' socio-economic groups corresponds to the pair of rows of Table 1 examined. For example, if we take the second and third rows of Table 6.1, (SEG's *5* and *10*), and index of dissimilarity is given by:

$$| 2.5 - 0.3 | + | 2.8 - 2.5 | + | 5.7 - 2.4 | +$$
$$| 14.7 - 7.2 | + | 27.9 - 4.1 | + | 7.4 - 5.7 | +$$
$$| 9.0 - 2.5 | + | 0.8 - 1.6 | + | 1.6 - 3.8 | +$$
$$| 17.2 - 47.0 | + | 3.3 - 12.0 | + | 6.6 - 9.9 | +$$
$$| 0.6 - 0.8 | = 90.2$$

Clearly, if the occupational destinies of sons with fathers in SEG's 5 and 10 had been the same (in a probabilistic sense), then the appropriate pair of rows of Table 6.1 would have had identical elements, and the index of dissimilarity would have taken its minimum value of zero. The maximum possible dissimilarity would have been obtained if there had been no overlap at all between the occupational destinations of fathers from SEG's 5 and 10. When divided by two, this index of dissimilarity can be interpreted as the percentage of sons of one or other father's category who would have to change their jobs, in order for the index to take its minimum value. Dividing 90.2 by two gives 45.1 as the index of dissimilarity between SEG's 5 and 10; i.e. 45.1 per cent of the sons of shopkeepers would have to shift their occupational destinations to different socio-economic groups, in order to make shopkeepers, (SEG 5) completely match skilled workers, (SEG 10) in terms of the occupational destinations of their sons. The index of dissimilarity has an interpretation, then. It is symmetrical, it cannot be less than zero, it is only zero when the two rows concerned are identical, and for any triple of rows, it satisfies the triangle inequality. Therefore it should behave like a distance between points in a metric space, and once all pairs of dissimilarities had been assembled, a *metric* MDS procedure in the classical manner might have been used to produce a solution for the thirteen SEG's of father's occupations in two or perhaps three dimensions. The differences between metric and non-metric modes of analysis in MDS is roughly as follows. In the former (metric MDS), the distances between pairs of points in the solution space are *linearly* related to dissimilarities between corresponding pairs of points in the data. In the latter case (non-metric) the distances between pairs of points in the solution space are ordinally related to dissimilarities between corresponding pairs of points in the data. As we have mentioned above, rank-order constraints in sufficient number appear to leave as little uncertainty about the relative positions of points in the solution space as metric constraints do. Macdonald chose to analyze the matrix of dissimilarity coefficients between socio-economic groups with a non-metric MDS procedure, (though he notes that analysis with metric MDS would have given very similar results). The coordinates of the thirteen socio-economic groups in a two dimensional solution (rotated to principal axes) are shown in Table 6.2, and plotted

Table 6.2. Coordinates of thirteen socio-economic groups in a two dimensional solution (rotated to principal axes) (Reproduced with permission from K. I. Macdonald, *The Analysis of Social Mobility: Methods and Approaches*, p. 210, Clarendon Press, Oxford, 1972)

	Dimension	
Socio-economic group	1	2
Farmers	−0.496	−1.944
Agricultural workers	−1.213	−0.453
Higher administrators, etc.	1.708	0.117
Other administrators, etc.	0.508	0.235
Shopkeepers	0.766	−0.516
Clerical workers	0.761	0.276
Shop assistants	0.297	0.220
Personal service	−0.120	−0.270
Foremen	−0.115	0.519
Skilled workers	−0.352	0.550
Semi-skilled workers	−0.592	0.547
Unskilled workers	−0.691	0.283
Armed forces (OR)	−0.461	0.534

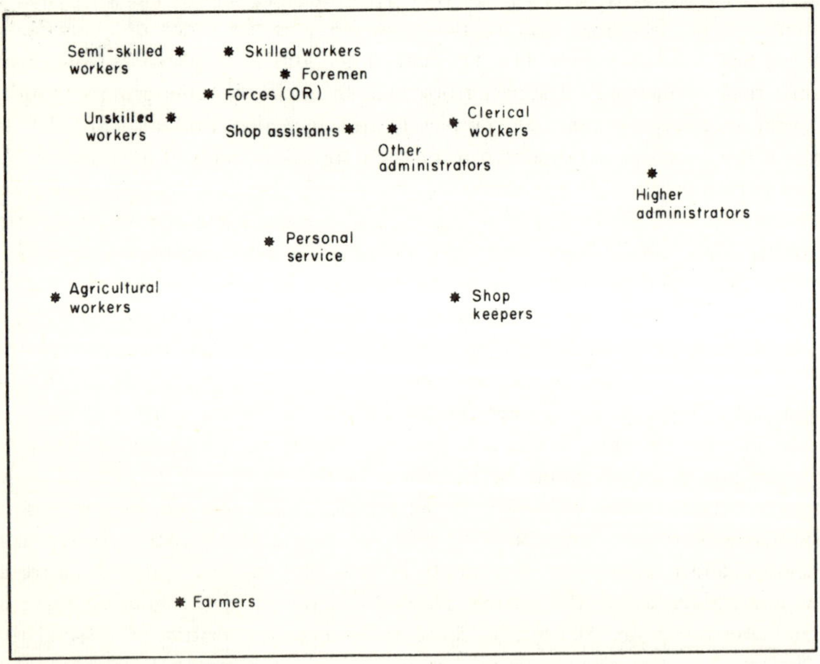

Figure 6.1. Plot of the two-dimensional solution for thirteen socio-economic groups, as given in Table 6.2 (Reproduced with permission from K. I. Macdonald, *The Analysis of Social Mobility: Methods and Approaches*, p.210, Clarendon Press, Oxford, 1972)

in Figure 6.1. As with factor analysis, the reference axes (dimensions) of any solution can be rigidly rotated without losing any information, though such a rotational transformation might be of considerable aid in interpreting the overall pattern of points in (as it happens here) a two dimensional space.

Another similarity between factor analysis and MDS relates to the problem of substantively 'interpreting' dimensions in terms of the projections of points upon them. Macdonald tentatively interprets the first principal axis, (drawn as the horizontal dimension of Figure 6.1), as, 'the educational base which an occupation provides its sons', and it should be readily apparent that the second dimension, (vertical in Figure 6.1) mostly serves the purpose of separating farmers (SEG 1) from other socio-economic groups. The nearer are any pair of socio-economic groups on the 'map' shown as Figure 6.1, the more similar are the occupational destinations of the sons of fathers in those socio-economic groups. With this interpretation in mind, the pattern of socio-economic groups shown in Figure 6.1 makes a certain amount of sense, and indeed it is precisely a picture of 'social distance' between socio-economic groups in a father–son mobility sense.

6.3 EXTENSIONS OF THE NON-METRIC MULTIDIMENSIONAL DISTANCE MODEL

The original non-metric MDS paradigm referred to the analysis of:

(i) *square*
(ii) *symmetric* data matrices of (dis)similarity coefficients by means of a
(iii) *Euclidean*
(iv) *distance* model involving
(v) a *monotonic transformation* of the data.

(In this framework, principal component analysis can be considered as the analysis of *a square symmetric* data matrix, but by means of a *vector* model involving a *linear* transformation of the data). Historically speaking, an important restriction has been that:

(vi) *analysis was traditionally confined to single 'two-way' data matrices.*

Each of these restrictions has been removed, allowing the extension of MDS to a very wide class of models for very different types of data, encompassing many situations encountered in survey research. These generalizations of the paradigm may now be taken up:

(i) and (ii) Perhaps the most useful generalization came in the extension from square, symmetric data to conditional similarity data, where ordinal constraints are to be satisfied conditionally, i.e. only within rows or only within columns. Data relating two distinct sets of objects (e.g. individuals and questionnaire items) can thus be analysed in the MDS framework, as can square *asymmetric* data matrices. It is very common for survey researchers to collect data of this first sort. For example, subjects may be asked to rate or rank occupations in terms of social usefulness, washing-powders in terms of cleansing power, role-expectations in terms of appropriateness, and so forth. An instance of the latter (square asymmetric matrices) would be sociometric rankings, where each member of a group judges all the other members of the group in terms of a criterion such as friendship.

A commonly used model for scaling such data is Coombs' unfolding model, which maps both a set of points representing stimuli (the objects of choice or preference) and a set of points representing subjects (where each point represents the most preferred or 'ideal' location of a subject) into a common space according to the rule that an individual's preference order is matched by the rank order of the distances between his (fixed) ideal point and the set of stimuli. Several such 'multidimensional unfolding' programs are available. An interesting application of the multidimensional unfolding model has been reported by Levine (1972), in a study of interlocking directorates which focussed upon the relationship between the boards of 14 banks and the boards of 96 industrial corporations. Making an analogy with the usual model of an individual's preference judgements over a set of stimuli, Levine chose to consider banks as 'individuals' and industrial corporations as the 'stimuli'. A bank was considered to be 'nearest' to the corporation with which it had most joint directors, and 'furthest from' the corporation with which it had the fewest number of joint directors. The data consisted then of 14 rank orders, each of the same 96 industrial corporations and in each of which there were a large number of tied ranks (where a bank had the same number of joint directors with corporation A as with corporation B). The solution provided an informative mapping of both banks and industrial corporations, represented as points in a common three-dimensional space.

(iii) To date, most MDS studies have used the *Euclidean* distance model, whether through convenience, beliefs about its robustness, or attachment to its substantive implications (Shepard, 1964; Sherman, 1972). However, other types of distance have been incorporated in the MDS paradigm (Carroll and Wish, 1975), and have been used in social science applications (Arnold, 1971).

(iv) The distance model has been extended to other models. Lingoes (1972) and others have developed non-metric analogues of both factor analysis and multiple regression, seeking a solution according to the assumptions of the appropriate linear model. Once again, the purpose is to provide the best solution to a monotone transform of the symmetric data matrix of (dis)similarities. An example of a MDS vector model for the analysis of individual preferences (analogous to Unfolding analysis) is provided by Carroll and Chang's (1964) MDPREF model (see also Kruskal and Shepard, 1974) where stimuli are represented as points in a multidimensional space, but each subject's set of preferences is represented in the stimulus space as a *vector*, or line directed towards his region of greatest preference. The order of projections of stimuli points on this vector represents the subject's rank order of preference. Starting either from a set of scale values obtained from pair comparison matrices or from rankings (together with a user-specified dimensionality) the model uses the Eckart-Young (1936) theorem to produce a solution where the subject's preference ordering is optimally matched by the order in which the stimuli project on the vector representing the subject.

In a study of subjective aspects of social stratification we have asked a number of subjects to rate a set of occupational titles in terms of what they consider is the degree of *social usefulness* of the occupation. The MDPREF model presented here provides one form of analysis suited to data such as these which are regularly collected in survey contexts. The two-dimensional MDPREF solution illustrates the utility of such a model, and is presented in Figure 6.2.

(In order to prevent the stimulus points from being obscured, each subject is represented only by the *terminus* of his vector from the origin of the space)

Whilst a detailed analysis would not be appropriate here, several comments are in order. The stimulus configuration is easily interpretable. A certain

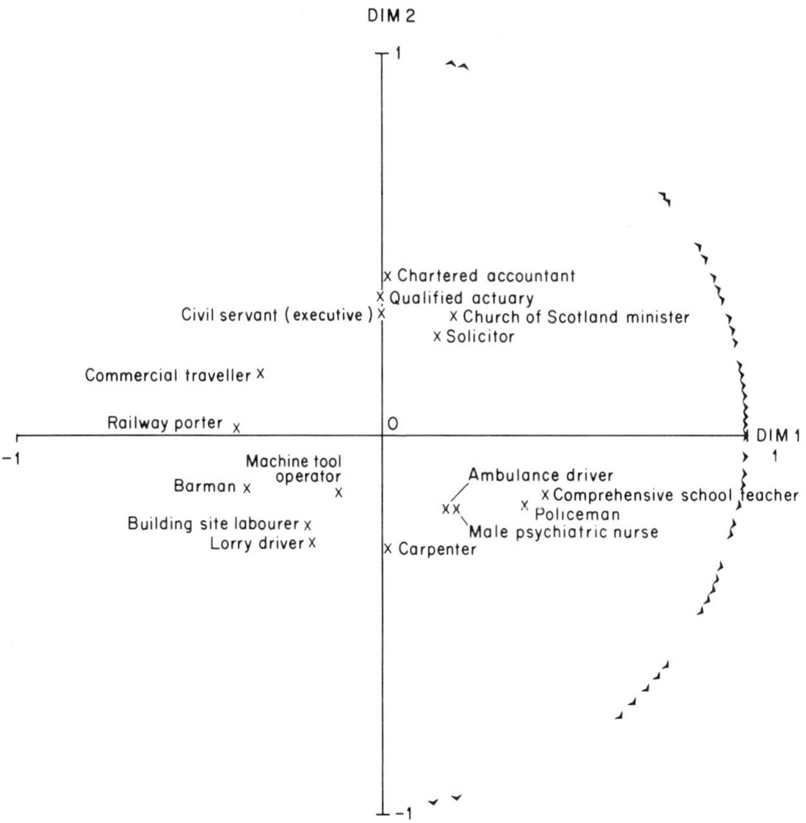

Figure 6.2. Rankings of social usefulness of 16 occupations. Joint space (two-dimensional MDPREF solution).

amount of consensus is evident in that the majority of the preferences are oriented in the direction of the horizontal axis of the figure, but systematically different points of view are also evident, and the range of preferences encompasses subjects with almost diametrically opposed rankings. It is important to bear in mind that this model provides an easily comprehensible graphic representation of individual differences in preference orderings.

(v) The first MDS models were designed to preserve metric information in the data, making the assumption that the empirical dissimilarities were a *linear* function of the model distances. Non-metric MDS models seek only to preserve ordinal information, and hence the much broader class of monotonic functions is permissible. In MDS procedures, the distinction between the two types of function has been defined by the form of regression used in obtaining a solution – in particular, linear regression of data upon distances is used in the metric case, whereas monotonic regression (Kruskal, 1964; Barlow *et al.* 1972) has been used in the non-metric case.

(vi) *Extensions of the MDS approach to Higher-Way Matrices.* We now turn to a discussion of the ways in which the MDS approach has been extended to 'higher-way' matrices. (By this is meant the simultaneous treatment of *a number* of two-way-(rows and columns)-matrices. Such data occur, for instance, in semantic differential research, where the basic data matrix is three-way: individuals by concepts by scales).

In the usual non-metric distance model the data (dissimilarities) are viewed as a *monotonic* function of the (Euclidean) distances between the points representing the stimuli:

$$\delta_{jk} = M(d_{jk}) = \left\{ \sum_{a=1}^{r} (x_{ja} - x_{ka})^2 \right\}^{\frac{1}{2}} \tag{1}$$

where δ_{jk} is the dissimilarity between points j and k, d_{jk} is the corresponding distance which may be represented in a r-dimensional Euclidean space, whose coordinates define a $\mathbf{X}$ space and M is the monotonic function relating the dissimilarities and the distances.

In the context of social surveys, we are often dealing with samples which include subgroups that are quite different in their characteristics, and different not only in their means and variances, but also in their patterns of covariance. If the subgroups are all fairly alike then it is reasonable to aggregate all the data and analyse them according to the distance model. But if the groups are quite different, and their similarity matrices are simply aggregated together, then this procedure is likely to wash out the systematic differences between subgroups. Another alternative might be to scale the data of each subgroup separately, but this leaves unanswered the question of how they might be compared to each other or referred to some common set of coordinates.

In a psychological context, this problem arises under the heading of individual differences in cognitive structure, and much of the methodological thinking in this area has been stimulated by this problem. As one way of approaching the problem of aggregation, Horan (1969) proposed a 'normal attribute space', or 'group stimulus space' defined by *all* the independent attributes or dimensions used by, and differentiating between, a set of perceivers. Any particular individual's (or subgroup's) 'private' cognitive space can then be thought of as a subspace of this Group attribute space, and those dimensions which a subject did not 'use' could be thought of as being given a zero weight by him. It is only a simple extension of this idea to think of subjects or subgroups as attaching *differential* 'importance', 'salience', or weight to each dimension of the space. Put slightly differently, each subject's data can be conceived as being the result of applying his own 'subjective metric' (systematic stretching or shrinking) to the dimensions of the Group Space.

In such a model, each individual can also be depicted in a 'Subject Space' (which has the same dimensions as the 'Group Space'), as a point located by the weights he has on each dimension. Then, by applying the individual's set of weights to the 'Group Space' – that is, rescaling it in the respondent's own subjective metric – the individual's own 'private space' is produced. This Individual Differences Scaling (INDSCAL) model was developed by Carroll and Chang (1970).

The main difference between the INDSCAL model and the simple distance model is that the subject's (dis)similarity judgment is represented as a *weighted* distance in the Group Space, and in this sense INDSCAL involves a simple generalization of the distance model. The dissimilarity between stimulus j and stimulus k for individual i is still assumed to be function (linear or monotonic, depending upon the investigator's choice of model) of distances in his own private space, which we may define by the matrix **Y**:

$$\delta_{jk}^{(i)} = F(d_{jk}^{(i)}) \qquad (2)$$

with

$$d_{jk}^{(i)} = \left\{ \sum_{a=1} (y_{ja}^{(i)} - y_{ka}^{(i)})^2 \right\}^{1/2} \qquad (3)$$

but these 'private' distances can equivalently be thought of a being modified, weighted distances in a common (Group) space, **X**:

The group (x_{ja}) and the private (y_{ja}) coordinates are related by:

$$y_{ja}^{(i)} = w_{ia}^{\;1/2} x_{ja} \qquad (4)$$

$$\delta_{jk}^{(i)} = F\left\{ \sum_{a=1}^{r} w_{ia}(x_{ja} - x_{ka})^2 \right\}^{1/2} \qquad (5)$$

Equation (2) states that for the ith subject the dissimilarities (δ_{jk}) between pairs of stimuli are some function, (linear of merely monotonic) of distances (d_{jk}) between the corresponding pairs of stimuli in that subject's solution space **Y**. The stimulus points in this space are referred to by their projections on a set of axes y_1, y_2, y_3 etc., and as equation (3) shows, the model assumes that distances between stimulus points, so far as the ith subject is concerned are obtained through applying the Euclidean distance function to the projections of the appropriate pair of stimuli upon the reference axes. The INDSCAL model was originally designed to represent the dissimilarity coefficient matrices of a number of individuals, each judging the same set of stimuli. Equation (4) shows how the reference axes in the private spaces $\mathbf{Y}^{(i)}$ are related to those of an overall group space, **X**.

The individual weights, w_{ia}, can be interpreted psychologically as the 'saliences' or 'importances' of dimension *a* to individual *i*. More generally, the incorporation of these weights into the model allows one of the individual matrices of pairwise distances between points to have systematically larger or smaller distances along some dimension than is the case for another of the individual matrices. Equation (5) puts the preceding equations together, and shows how the *i*th individual's judgment of the dissimilarity between stimulus *j* and stimulus *k* is a function, *F*, of the idiosyncratically weighted Euclidean distance between those points when they are referred to by their projections on the axes x_1, x_2, x_3 etc. of the common group space, **x**.

6.4 WAYS IN WHICH INDSCAL CAN BE APPLIED

The most direct application of INDSCAL is the anslyis of the pairwise similarity judgments given by a set of subjects. Such data are not normally collected in conventional social surveys, though there is little reason whey they should not be (for examples see Wish *et al.* 1970; Coxon and Jones, 1974). However it is possible to conceive of similarities data as coming from sources other than individual subjects, and the term 'pseudo subject' is often used to cover this extended sense. In the example we give below, *groups* of individuals serve as 'pseudo subjects' and distances are derived from correlation coefficients.

Survey analysts often make preliminary explorations of their data by inspecting the square symmetrical matrix of coefficients of association between all pairs of relevant variables. With the aid of such inspection, they make decisions about what variables they will combine into composite indices of one kind or another. It is very common for analysts to use some technique such as principal component analysis, factor analysis, or cluster analysis to aid such inspection (see Chapters 3 and 4). Non-metric MDS can be regarded as another tool in this particular data-reduction kit, and it has the advantage that only ordinal assumptions need to be made of the measures of association between variables. For example Napior (1972) has given an expository analysis of a matrix of Goodman–Kruskal gamma coefficients between sixteen variables using three non-metric models: a hierarchical clustering algorithm, Kruskal's MDSCAL and the Guttman–Lingoes non-metric factor analysis. These analyses resulted in the identification of several unidimensional sets of items which were then separately scaled using a simple Guttman scaling procedure. MacRae (1970) has worked in similar fashion with legislative voting data. He computed Yule's *Q* as an index of similarity between each pair of bills for which representatives could vote in favour of or against, and obtained a two dimensional representation of the resulting similarities matrix using non-metric MDS.

But whatever technique is used to aid the naked eye of the researcher, so long as he examines his data in an aggregated form, he still runs the risk of putting together

Table 6.3. Self-rating type items dealing with 'Occupational Values' as administered to a representative sample of able school leavers in Scotland

1.		I have given a good deal of thought to the level of income I hope to be earning in my late twenties. (INCOME). (9-point scale from 'strongly agree' to 'strongly disagree')
2.		. . . opportunity to work with people rather than things. (PEOPLE). ('My ideal job would . . .'; 9-point scale from 'absolutely essential' to 'definitely not wanted').
3.		. . . opportunity to hold a senior position by the age of 30. (PROMOTION).
4.		. . . opportunity to follow my academic and intellectual interests full-time. (ACADEME).
5.		. . . enable me to look forward to a stable secure future. (SECURITY).
6.		. . . opportunity to help young people. (HELP).
7.		. . . leave me relatively free of supervision by others. (FREE).
8.		. . . require me to produce original and creative work. (CREATE).
9.		. . . excitement or adventure. (EXCITE).
10.	(A)	. . . absorb my energy and interests both inside and outside working hours. (ABSORB).
11.	(B)	. . . chance to exercise moral leadership (LEADERSHIP).
12.	(C)	. . . involve travelling in the United Kingdom and abroad. (TRAVEL).
13.	(D)	. . . work full-time with ideas and theories. (IDEAS).
14.	(E)	. . . opportunity to work with young children. (CHILDREN).
15.	(F)	. . . the power to make important decisions. (POWER).
16.	(G)	. . . a clear cut work-routine. (CLEARCUT).
17.	(H)	. . . enable me to move up in the world. (ASCEND).
18.	(I)	. . . allow me to stay near parents and family. (PARENTS).
19.	(J)	. . . not commit me for long to any one type of work. (NOT COMMIT).
20.	(K)	. . . enable me to freelance or be my own employer. (FREELANCE).
21.	(L)	How important do you think it is that a job should provide the chance to earn a lot of money? (MONEY. (9-point scale from 'extremely important' to 'not at all important').

subgroups of cases with differing patterns of relationships between variables Psychologists have expressed this notion by suggesting that different groups of persons may show different factor structures over the same battery of tests. The following example illustrates how the Carroll–Chang INDSCAL model gives the survey researcher some leverage on this problem, and in addition, provides a simplified visual presentation of a mass of complex numerical data.

The data for this example consisted of responses to a short inventory of 21 items which were concerned with the values that people might satisfy in work. The subjects who completed the inventory were a representative sample of educationally able Scots who left school in 1970 (see Jones and McPherson, 1972, for details of sampling).

It was reasoned *a priori* that differences in the statistical patterning of relationships among occupational values items might well be related to differences in the average profiles of such values between diverse groups. Therefore six sub-groups from the total sample were defined as being likely to have markedly different value profiles. These groups were:

1. Men who entered upon Engineering or Technical courses in Universities. ($n = 93$)
2. Men who entered upon Science courses in Universities. ($n = 165$)

3. Men who entered full-time employment, with no part-time education or training. ($n = 175$)
4. Women who entered upon Language or Literature courses in Universities. ($n = 115$)
5. Women who entered upon the Primary School Teaching course in College of Education. ($n = 224$)
6. Women who entered full-time employment with no part-time education or training. ($n = 195$)

University course groups were defined in accordance with the conventions used by the Universities Central Council on Admissions. Product-moment correlations were calculated between each pair of the 21 values items within each of the six sub-groups.

The six correlation matrices might perhaps be interpreted as being matrices of similarity coefficients between values. The larger the correlation between two variables, the greater the similarity between them, and the 'closer' they are in a space where the variables are represented as points. However once we have 'bought information' by assuming linearity in relationships, which are necessary in the calculation of correlation coefficients, it makes little sense to pretend that we do not have it. It is easy to show that distances are simply related to covariances and Pearsonian correlations (scalar products) (see Torgerson, 1958 or van de Geer, 1971), and may be simply converted. For covariances,

$$d_{ij} = (s_i^2 + s_j^2 - 2s_is_jr_{ij})^{1/2}$$

where s_i^2 is the variance of variable i, and r_{ij} is the product moment correlation between variables i and j.

In the case of normalized (unit-length) variables, this equation reduces to

$$d_{ij} = (2 - 2r_{ij})^{1/2} = 2(1 - r_{ij})^{1/2}$$

and the interitem distances may be estimated from interitem correlations either with or without taking account of the standard deviations of the items concerned. Clearly the research worker must make his own decision about which method to use in a case like this. Since it is not normal practice to standardize variables *within* sub-groups before forming scales or analysing the data, we favour the use of distances as estimated from covariances (and standard deviations) and it is this analysis that we report here. The six matrices of inter-item distances were fitted to the INDSCAL model which has been described above. The metric version of INDSCAL was used, as we wished distances in the solution space to be a linear function of ('distances' in) the data.

Scaling was carried out in three dimensions and also in two dimensions, and the coordinates of values in the group-space configurations are shown in Table 6.4. It should be borne in mind that the (correlated) dimensions of the group-space of an INDSCAL solution are 'fixed' in the sense that any rotation would destroy the statistical (least squares) properties of the solution. It therefore makes sense to try to interpret these axes, which may tentatively be labelled in terms of the items which have extreme projections on them.

Table 6.4. Three- and two-dimensional co-ordinates of the INDSCAL solution, (successive solution, IOY = 1) of matrices of product-moment correlations between 21 questionnaire items, (interpreted as indices of similarity between the items).

		INDSCAL coordinates				
		3-D Solution			(IOY = 1) 2-D Solution	
Abbreviated questionnaire		I	II	III :	I	II
1.	INCOME	0.283	0.054	−0.062	0.283	−0.053
2.	PEOPLE	−0.081	−0.283	0.202	−0.081	0.283
3.	PROMOTION	0.189	0.176	−0.132	0.189	−0.176
4.	ACADEME	−0.054	−0.022	−0.330	−0.054	0.022
5.	SECURITY	0.375	−0.128	−0.038	0.375	0.128
6.	HELP	−0.093	−0.369	0.176	−0.093	0.369
7.	FREE	−0.125	0.090	0.015	−0.125	−0.091
8.	CREATE	−0.288	0.027	−0.240	−0.288	−0.027
9.	EXCITE	−0.268	0.229	0.025	−0.268	−0.229
10. (A)	ABSORB	−0.276	−0.134	−0.187	−0.276	0.134
11. (B)	LEADERSHIP	−0.077	−0.210	−0.214	−0.077	0.210
12. (C)	TRAVEL	−0.168	0.338	0.019	−0.167	−0.338
13. (D)	IDEAS	−0.149	−0.069	−0.318	−0.149	0.069
14. (E)	CHILDREN	−0.158	−0.376	0.397	−0.158	0.376
15. (F)	POWER	0.039	0.078	−0.244	0.039	−0.078
16. (G)	CLEARCUT	0.342	−0.218	0.078	0.342	0.219
17. (H)	ASCEND	0.264	0.222	−0.142	0.264	−0.222
18. (I)	PARENTS	0.229	−0.179	0.253	0.229	0.179
19. (J)	NOT COMMIT	−0.077	0.303	0.435	−0.077	−0.303
20. (K)	FREELANCE	−0.225	0.306	0.230	−0.224	−0.306
21. (L)	MONEY	0.319	0.166	0.078	0.319	−0.166

Taking the two-dimensional solution only (since the fit for this model is only slightly worse than the fit for the 3-dimensional model), the first dimension separates items to do with financial well-being and the extrinsic rewards of a job from those involving intrinsic work satisfactions, (INCOME, MONEY, CLEARCUT, SECURITY, ASCEND *versus* CREATE, ABSORB); the second dimension, which might be labelled in terms of localist *versus* cosmopolitan values, differentiates items referring to work involving a certain kind of nurturant orientation towards children from those which might describe jobs giving a great deal of 'free-wheeling' variety, (HELP, CHILDREN, CLEARCUT, PARENTS, LEADERSHIP, *versus* EXCITEMENT, TRAVEL, NOT COMMIT, FREELANCE).

The configuration of values items is plotted in the two dimensional INDSCAL solution, in Figure 6.3. Clusters of similar items can readily be identified by eye, as for example (a) CHILDREN, PEOPLE, HELP, ABSORB, (b) PROMOTION, MONEY, POWER, INCOME, ASCEND, (c) EXCITE, FREELANCE, TRAVEL, NOT COMMIT.

The exploratory and descriptive analysis might cease at this point in a traditional study, but the INDSCAL model allows one to go further. Specifically, the investigator may ask about how well the overall configuration represents the patterns of relationship among occupational values in different subgroups. As was

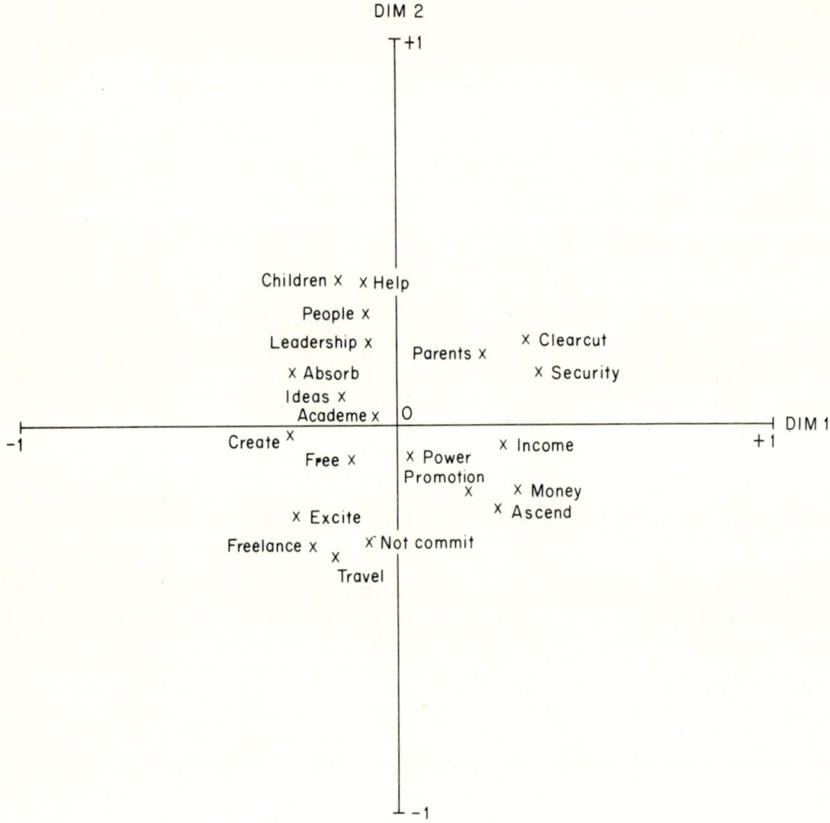

Figure 6.3. INDSCAL Group Space.

mentioned above, the model permits individual 'subjects' or groups to be represented as a transformation of the overall configuration, this transformation being accomplished by allowing each 'individual' uniformly to rescale ('stretch' or 'shrink') *each* dimension of the overall stimulus configuration by a multiplying weight. The appropriate weights for each of the six sub-group 'pseudo-subjects' are shown in Table 6.5, and the weights of the sub-groups are plotted in a two-dimensional 'subject space' in figure 6.4. The differences are interesting. On the whole, the male sub-groups are different from the female ones. The patterns of relationships between occupational values which are shown by men indicate that items are predominantly organized according to a factor of intrinsic versus extrinsic rewards, rather than in terms of localism – cosmopolitanism. In terms of 'private distances' between items for the male groups, less importance is attached to the second dimension than to the first. Female subgroups on the other hand appear to organize the items mostly through the factor of localism versus cosmopolitanism, though it should be borne in mind that what we have termed 'localism' here is bound up with sex-role-linked occupational values, such as helping young children.

Table 6.5. INDSCAL weights expressing the appropriate transformations of the overall 'compromise' configuration (see previous table) for six groups.

	3-D Solution			2-D Solution	
Group	I	II	III	I	II
1. Men engineering students	0.658	0.284	0.418	0.658	0.284
2. Men science students	0.588	0.437	0.359	0.588	0.437
3. Men employment	0.805	0.272	0.220	0.805	0.272
4. Women languages students	0.452	0.683	0.230	0.452	0.683
5. Women teacher-training	0.336	0.806	0.156	0.335	0.806
6. Women employment	0.739	0.394	0.183	0.739	0.394
Sums of squares	2.289	1.616	0.463	2.289	1.616

Goodness of Fit Indicator	3-D Solution	2-D Solution
Percentage of variance accounted for by INDSCAL solution	72.8%	65.1%

Note: The sequential solution was used, and axes are therefore orthogonal. (In fact I and II are correlated at 0.049 in both solutions, I and II are correlated at 0.015, and II and III at 0.084).

In the case of female groups, less importance is attached to the first dimension than to the second. It also seems that within each sex, the subgroups whose members entered full-time employment rather than higher education tend to organize the values items according to dimension I (extrinsic–intrinsic job rewards) rather than dimension II (localism–cosmopolitanism). One can of course use Figure 6.4 to compare the subgroups in terms of their relative patterning of weights. It is evident from this that women who entered full-time employment have a pattern of INDSCAL weights rather more like the pattern for male subgroups than like the pattern for women who continued into higher education.

The consequences of applying individual 'pseudo-subject' weights to the overall configuration can best be seen by contrasting the 'private spaces' for two extreme sub-group 'pseudo-subjects' of (a) Group 5: women studying for the primary teaching qualification and (b) Group 3: men who entered employment with no part-time education.

Clearly, the uniform stretching and shrinkings which the overall space has undergone in being transformed to the 'private space' for these two extreme groups indicate a plausible contrast in the patterns of relationship among the occupational values items. Among women who entered the primary school teaching course, the first (horizontal in Figure 6.5) dimension does not discriminate between items very much. The second (vertical) dimension continues to separate items along a continuum of their reference to helping people on a continuous and long term basis. The picture shown in Figure 6.5A contrasts dramatically with Figure 6.5B. There, the second (vertical in Figure 6.5B) dimension has been shrunk, and it is the

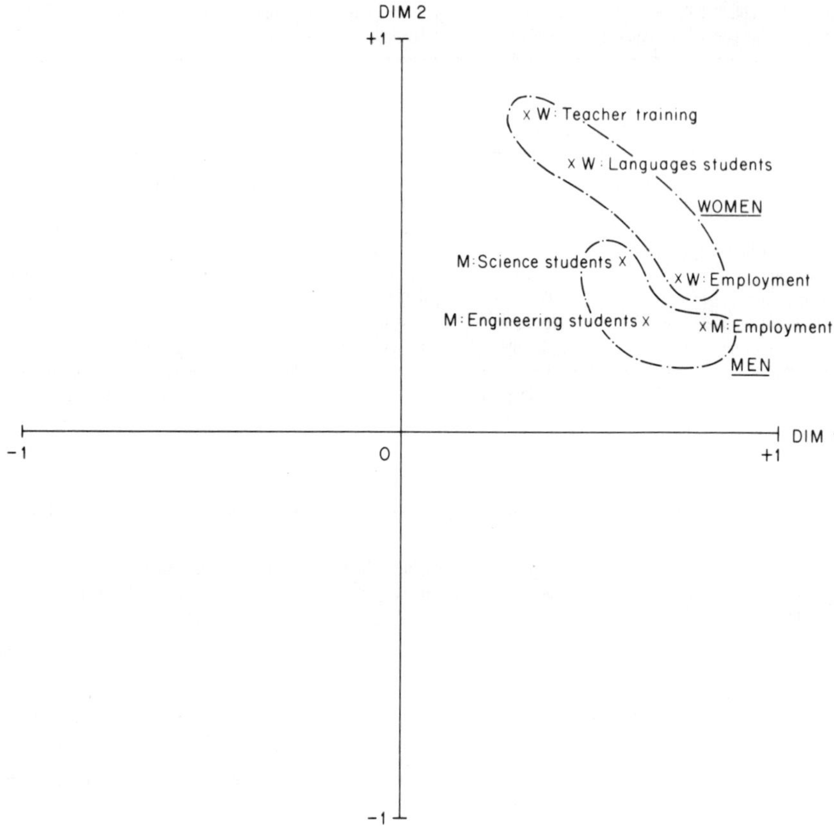

Figure 6.4. INDSCAL Subject Space.

first (horizontal) dimension which continues to discriminate between items, this time on the basis of their reference to internal or external systems of reward.

6.5 CONCLUSIONS AND DEVELOPMENTS

This application of the INDSCAL model is useful as a means of summarizing and presenting in the compact form the sort of correlational information that is often generated in the preliminary analysis of social survey data. It also provides an approach to one aspect of the 'factor matching' problem referred to earlier: the problem of deciding whether or not different samples of subjects who have all been measured on the same set of variables can be described by the same set of underlying factors. Similar issues are involved in describing and assessing concept-scale interactions in the semantic differential (see Bynner and Romney (1972), Jones (1972) and Coxon (1972) for further discussion). However, it is a major restriction of the INDSCAL model that the only legitimate transformation is that of differentially stretching or shrinking the fixed dimensions of the overall

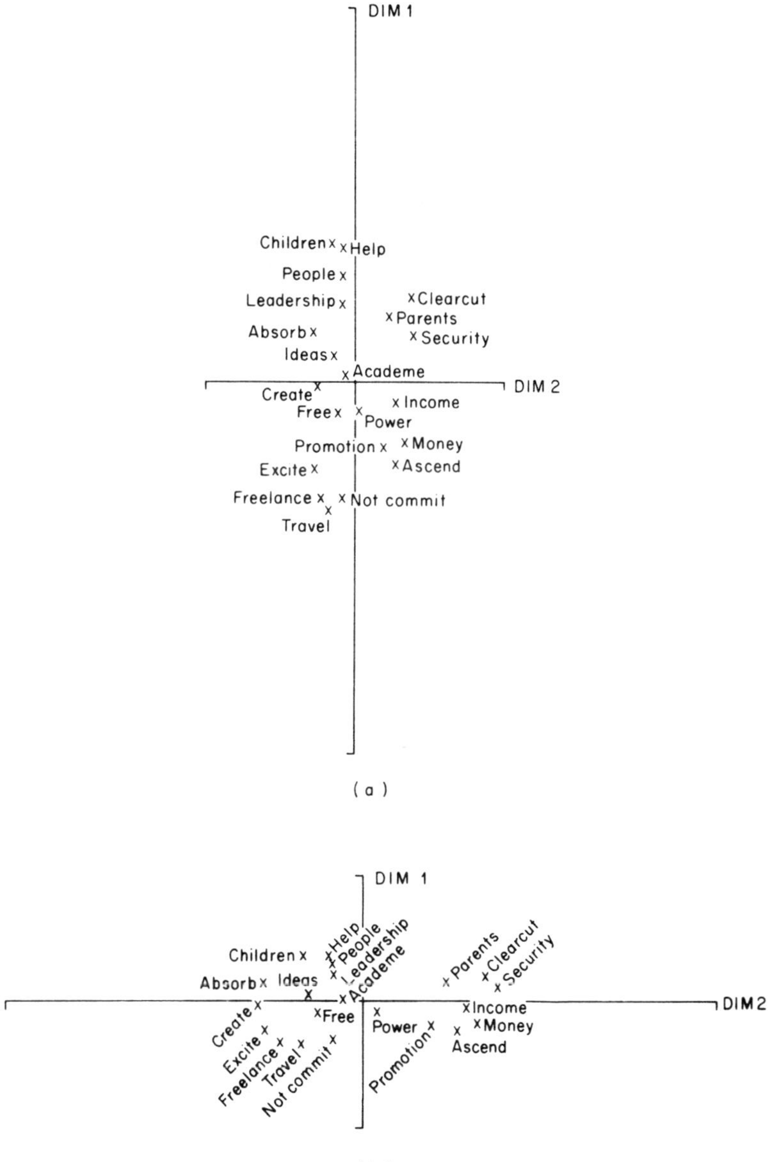

Figure 6.5. INDSCAL: Two 'Private Spaces':
(a) Group 5: Women, primary teaching students. Weights: 0.34, 0.91.
(b) Group 3: Men, no further education. Weights: 0.81, 0.27.

configuration. Rotation of axes is not permitted in the INDSCAL model, and the reason for this is as follows. If all subjects perceive an identical Euclidean stimulus space, then any rigid rotation is permissible, as it leaves distances unchanged. The INDSCAL model by contrast, asserts that whilst the distances in each *private* space are Euclidean, these distances are *not* Euclidean by reference to the Group Space, since they receive idiosyncratic weighting factors for each individual's own 'subjective metric', (see equation (5) above).

Any rotation of the group coordinates $\{x_{ja}\}$ changes each individual's 'subjective metric'. This means that the values predicted for the data by a fitted model whose group coordinates have been rotated, (for ease of interpretability or whatever) will always be a worse fit to the original data than was the case with the unrotated solution. The price of rotation is less good fit to the data.

Carroll and Chang (1970) have suggested that this property of the INDSCAL model brings the added advantage of direct interpretability. Certainly, many studies published by psychologists seem to support this claim, but it would be difficult to argue that substantive interpretability is directly linked to formal properties. Schönemann and Wang (1972, p. 443) makes the analogy with principal components:

> 'The fact that principal components are rendered unique by mathematically convenient constraint does not necessarily mean that they are also psychologically meaningful or useful'.

The uniqueness of the orientation of INDSCAL dimensions can be a substantive inconvenience, when a rotation of a Group Space configuration would provide a more interpretable reference frame; a *separate* INDSCAL analysis of the data of distinctive subgroups often reveals considerable differences in orientation of the axes. If such data are aggregated, the orientation of the overall group dimensions will be in large part, a compromise (in a bad sense) between the different orientations. A model which allowed differential rotation of axes, as well as their differential weighting would be a distinct advantage in such cases, for it would be possible to represent subjects using different combinations of the reference axes in their judgments. Carroll has outlined such a model, which is clearly a generalization of INDSCAL, but for most applications in survey analysis, developments such as this are likely to be of little more than academic interest. The importance of such models is that they show rather well that if the problem of aggregation in heterogeneous populations is to be taken seriously, then our models of analysis will have to match the complexity of the data.

REFERENCES

Arnold, J. B. (1971), 'A multidimensional scaling study of semantic distance', *J. Exptl. Psychol.*, **90**, 349–372.

Barlow, R. E. *et al.* (eds) (1972), *Statistical inference under order restrictions*, London: Wiley.

Benjamin, B. (1958), 'Inter-generation differences in occupation', *Population Studies,* **II**, 262–268.

Bynner, J. and Romney, D. (1972), 'A method for overcoming the problem of concept-scale interaction in semantic differential research', *Br. J. Psychol.*, **63**, 229–234.

Carroll, J. D. (1972), 'Individual differences and multidimensional scaling', In R. A. Shepard, A. K. Romney and Sara Beth Nerlove (eds), *Multidimensional Scaling: theory and applications in the behavioral sciences*, Volume I, New York: Seminar Press.

Carroll, J. D. and Chang, J. J. (1964), 'Nonmetric multidimensional analysis of paired comparisons data', Murray Hill, N.J.: Bell Laboratories *mimeo*.

Carroll, J. D. and Chang, J. J. (1970), 'Analysis of individual differences in multidimensional scaling via an *n*-way generalization of "Eckart-Young" decomposition', *Psychometrika,* **35**, 283–319.

Carroll, J. D. and Wish, M. (1975), 'Multidimensional perceptual models and measurement methods', in E. C. Carterette and M. P. Friedman (eds) (1975) *Handbook of Perception*, New York: Academic Press.

Coombs, C. H. (1964), *A theory of data,* New York: Wiley.

Cormack, R. M. (1971), 'A review of classification', *J. Royal Stat. Soc. (A)*, **134**, 321–367.

Coxon, A. P. M. (1972), 'Reply to Jones', *Sociology,* **6**, 453–455.

Coxon, A. P. M. and Jones, C. L. (1974), 'Occupational similarities: some subjective aspects of social stratification', *Quality and Quantity,* **8**, 139–157.

Eckart, C. and Young, G. (1936), 'Approximation of one matrix by another of lower rank', *Psychometrika,* **1**, 211–218.

Eisler, H. and Roskam, E. E. (1973), 'Multidimensional similarity: an experimental study of vector, distance and set theoretical models', Nijmegen: Dept. of Psychology, *mimeo*.

Hodson, R. R., Kendall, D. G., and Tautu, P. (1971), *Mathematics in the Archaelolgical and Historical Sciences*, Edinburgh: University Press.

Horan, C. B. (1969), 'Multidimensional scaling: combining observations when individuals have different perceptual structures', *Psychometrika,* **34**, 139–165.

Jardine, N. and Sibson, R. (1971), *Mathematical Taxonomy*, London: Wiley.

Jones, C. L. (1972), 'On occupational attributes', *Sociology*, **6**, 451–452.

Jones, C. L., and McPherson, A. F. (1972), 'Implications of non-response to postal surveys', *Scot. Ed. Studies,* **4**, 143–151.

Kruskal, J. B. (1964), 'Multidimensional scaling by optimizing goodness of fit to a nonmetric hypothesis', *Psychometrika,* **29**, 1–27.

Kruskal, J. B. (1964), 'Nonmetric multidimensional scaling: a numerical method', *Psychometrika*, **29**, 115–129.

Kruskal, J. B. (1966), 'Geometric interpretation of diagnostic data for a digital machine', *Bell Syst. Tech. J.*, **45**, 1299–1338.

Kruskal, J. B., and Shepard, R. N. (1974), 'A nonmetric variety of linear factor analysis', *Psychometrika,* **39**, 123–157.

Levine, J. H. (1972), 'The sphere of influence', *American Sociological Review,* **37**, 14–27.

Lingoes, J. C. (1972), 'A general survey of the Guttman–Lingoes nonmetric program series', in Shepard *et al.* (1972). vol. 1.

Macdonald, K. I. (1972), MDSCAL and distances between socioeconomic groups. In K. Hope (ed.), *The Analysis of Social Mobility: Methods and Approaches*, pp. 211–234, Oxford: Clarendon Press.

MacRae, D. Jr. (1970), *Issues and Parties in Legislative Voting*, New York: Harper and Row.

Napior, D. (1972), 'Nonmetric multidimensional techniques for summated ratings', in Shepard *et al.* (1972), vol. 1.

Schönemann, P. H. and Wang, M. M. (1972), 'An individual difference model for the multidimensional analysis of preference data', *Psychometrika,* **37**, 275–309.

Shepard, R. N. (1962), 'The analysis of proximities: multidimensional scaling with an unknown distance function', *Psychometrika,* **27**, 125–139 and 219–246.

Shepard, R. N. (1964), 'On subjectively optimum selections among multiattribute alternatives, in M. W. Shelley and G. L. Bryan, *Human Judgments and Optimality,* New York: Wiley.

Shepard, R. N. (1966), 'Metric structures in ordinal data', *J. Math. Psychol.*, **3**, 287–315.

Shepard, R. N., Romney, A. K., and Nerlove, S. B. (1972), *Multidimensional Scaling: Theory and Applications in the Behavioral Sciences*, Vol. 1, New York: Academic Press.

Sherman, C. R. (1972), 'Nonmetric multidimensional scaling: a Monte Carlo study of the basic parameters', *Psychometrika,* **37**, 323–355.

Sibson, R. (1972), 'Order-invariant methods for data analysis', *J. Royal Stat. Soc. (B)*, **34**, 311–349.

Sokal, R. R., and Sneath, P. H. A. (1963), *Principles of Numerical Taxonomy,* London: Freeman.

Torgerson, W. S. (1958), *Theory and Methods of Scaling*, London: Wiley.
van de Geer, (1971), *Introduction to Multivariate Analysis for the Social Sciences*, London: Freeman.
Wish, M., Deutsch, M., and Biener, L. (1970), 'Differences in conceptual structures of nations': an exploratory study', *J. Pers. and Soc. Psychol.*, **16**, 361–373.

Chapter 7

Attitude Scale Construction

A. C. McKennell

7.1 INTRODUCTION

Much of the impetus towards attitude scale construction stems from the well-established limitations of single question opinion gauging. The ease with which large variations in percentages can be produced by varying the question format led, a generation ago, to a literature replete with studies serving merely to demonstrate this fact. In what still remains the most comprehensive review of these early studies, McNemar (1946) noted (and it remains true today) that 'so much is known about the variations which can be produced, and so little is known about which variation is most nearly correct'. He recommended accordingly that 'single question opinion gauging be discarded in favour of opinion measurement by attitude scales'. Eysenck (1954, Chapters 2 and 3) in providing a convenient summary of much the same literature, echoed this conclusion, and further emphasized the advantages of multi-item scales. In view of the fact that opinion pollers, and even some sociologists, continue to report percentage replies to single items as though such figures had absolute meaning, it is perhaps timely that Schuman and Duncan (1974) have provided a more recent reminder that not only response distributions but 'one's conclusions about the most basic background correlates of an attitude can vary decisively with item wording and format'.

Despite its current neglect by some practitioners the research noted above established beyond doubt the serious deficiencies of single items and the superiority of multi-item scales in respect of such matters as reliability, unidimensionality, and freedom from specific wording bias. From the standpoint of this chapter we would emphasize particularly how use of the single item can by-pass what is important to respondents, imposing on them a biased frame of reference different from that within which they normally think about the attitude object or issue. To take a fairly recent, if extreme example. As part of a larger study, Goldthorpe *et al.* (1968) used the following question to assess the attitudes of car workers to management-worker relations. 'Here are two opposing views about industry generally: I would like you to tell me which you agree with more. Some people say that a firm is like a football side – because team work means success and is to everyone's advantage. Others say that team work in industry is impossible – because employers and men are really on opposite sides. Which view do you agree with more?' Put like that, more than two thirds took the view that team work was to everyone's advantage. This result was consistent with the hitherto peaceful history of labour relations in the plants where the interviews had been carried out. The authors therefore concluded that the workers had a 'harmonistic image of the enterprise'. But it happened that the research was conducted at the very time when the era of calm labour relations was about to come to an abrupt end. Even before the results could be published the inadequacies of the attitude measurement method stood revealed in the form of outbreaks of unrest and a succession of labour disputes in the plants where the interviewing had been done. We do not wish to raise here the vexatious issue of the prediction of behaviour from attitudes (on which there is now a large and still growing literature; see Lemon (1973), Schumon (1973), Brannon *et al* (1973), and the comments in McKennell (1974, pp. 28–29)). But as a bare minimum, whether the interest is in predicting behaviour or simply in unbiased measures of attitudes, it seems obvious that the method used should tap what is important and representative in the respondents' outlook. A theme of this chapter is that while a single item phrased at the desk can hardly do this, even several items can fail without adequate qualitative work at the item elicitation stage. On the other hand, once the important dimensions have been tapped by such exploration, the selection and use of even one item to represent each dimension can be justified if that item's psychometric properties have first been established at the scale development stage. For most attitude dimensions, however, as will be seen, the quest for adequate reliability of measurement will require several items and hence the construction and use of an attitude scale.

Given the limitations of the single item, techniques are required for selecting several items and for combining item responses to yield an overall score for individuals which will reflect their positions on the underlying attitude dimension with adequate reliability. Most longstanding among the alternative rationales and operational routines for doing this are the conventional Thurstone, Likert and Guttman scaling models. Good accounts of the standard, unidimensional scaling procedures can be found, for example, in Upshaw (1968), Moser and Kalton (1971), Lemon (1973). They are largely substitutable in that for those attitude continuums that can be and have been measured by all three methods, the rank ordering of respondents is usually very close whichever of the three methods is applied. (See Seiler and Hough, 1970.) Compendiums of ready-made attitude scales, usually constructed by one of the three classical methods, are available covering a wide variety of topic areas; see Robinson *et al.* (1968–69), Shaw and Wright (1967), Miller (1969), Bonjean *et al.* (1968).

The classical scaling procedures represented an important step forward at the time of their development, and remain all that is required if the researcher can be sure that his purposes will be met by scaling attitudes along one, single dimension. As the goals of an attitude survey become more thorough-going and ambitious, however, a single scale, either ready-made or specially constructed, becomes increasingly inappropriate as a representation of the structure of the attitudes under investigation. The ideal yield of any detailed survey of the attitudes of a large cross-sectional sample should be an analysis and interpretation of the pattern of interrelationships that hold between the attitudinal data and various specific background and behavioural variables. Even where the assumption of a single general factor underlying the attitudinal domain holds good, in the special sense discussed below, the results of measuring evaluative dispositions only at this most general level can be too crude to be useful at the analytical stage. More particularized dimensions or components need to be taken into account, and techniques become necessary which provide scope not only for the measurement but also for the exploration and discovery of the dimensions along which individuals should be compared. Moreover, if several dimensions are to be measured along with non-attitudinal data, then to prevent the final questionnaire from becoming inordinately long the scaling method should provide a rationale for economizing in the use of items: in particular, in addition to a procedure for item analysis, it should provide information on the cost in reliability when a reduced set of items is carried forward from pilot to main survey.

The above features of a thorough-going exploratory attitude survey will be emphasized in this chapter. Since there is by now an abundance of good texts to which one can turn for details of the standard unidimensional scaling procedures, only a summary account of these will be presented here. Space will be devoted instead to what is really a multidimensional extension of principles that are implicit, if not always brought out, in the original Likert or summative scaling procedure. This extension is possible because the Likert procedure is a special case of the psychometric test model, in the theory of mental tests, as described in Lord and Novick (1968), for example. The emphasis will be on those aspects of the psychometric test model that render it especially suitable for the scale development stage of an exploratory attitude survey. It will be convenient to refer to the more general approach as the 'alpha' technique, after the alpha coefficient of reliability which is central to the method. Once the basic ideas are grasped the method is simple to apply. Its relation to other techniques, both older and more modern, is developed throughout the chapter, and particularly towards the end.

7.2 THE STANDARD UNIDIMENSIONAL SCALING PROCEDURES

The three classical scaling models are all based on the notion of a quantitative continuum along which people can be arrayed, depending on the magnitude of the attitude which they possess. What distinguishes the Thurstone and Guttman methods from the Likert procedure is that in the former models the items are also assigned to positions along the continuum. In the Thurstone method item magnitudes are assessed through the operations of a group of judges, hence it is also known as the Judgement method. Thurstone (1928) recommended collecting an initial item set of between 100 and 150 statements culled from a wide variety of published source material on the object under study. These are then given to about 100 judges who are instructed to sort them into some eleven categories in terms of their favourability or unfavourability to the object. The method is sometimes referred to as the method of 'equal appearing intervals' on the (unvalidated) assumption that the intervals between the categories are in fact equal. After sorting by the judges, items with a small interquartile range are retained as being less ambiguous, and the median position of each retained item is taken to indicate its scale value (sometimes referred to as consensual location scoring). The finished scale would consist of twenty or so items selected to give adequate coverage across the whole range of scale positions. In applying the scale respondents would be scored according to the mean or median scale value of the items with which they agree. An unresolved problem for this scaling procedure is that the original judges may not be able to make their assessments independently of their own attitudes (Hovland and Sherif, 1952). Another is the difficulty of formulating statements that have the (nonmonotone) characteristics to justify consensual location scoring; that is statements which will not be endorsed by respondents whose attitudes are either more or less favourable than the scale position of the item. Further points comparing the Thurstone method with the alpha approach are made towards the end of the chapter.

The scaling theory developed by Guttman (1944, 1950) is based on the assumption that a set of items which measure a single unidimensional trait can be ordered along a continuum of 'difficulty', the order being reflected by the relative infrequency with which each item is endorsed. Items of this type are said to be cumulative in that acceptance of one implies that the person possesses enough of the attribute to accept all items of a lesser magnitude. An example is a scale used by the author (McKennell, 1969) to measure Readiness to Complain (about aircraft noise) in which the items, in cumulative order, were Willingness to Visit an Official, to Write to an Official, to Attend a Meeting, and to Sign a Petition. In a perfect cumulative scale if one knows a respondent's total score (how many items he accepts) one knows exactly which items have been accepted and which have been rejected. A respondent's answer pattern for a set of items is then said to be completely 'reproducible' from his total score on the set. On the strict application of Guttman scaling, as he initially set out the method, the primary task of the scale constructor is to determine whether or not the attitude area in question is scalable according to the cumulative model. This is done by first intuitively formulating a relatively small number of items (Guttman recommended at least ten items but most published scales are established on fewer) the content of which appears on *a priori* grounds to relate to the attitudinal concept under investigation. The answer patterns for a small sample of respondents, usually one hundred, are then obtained and examined in detail to see how far they approximate the cumulative model. There is a considerable literature and some dispute on what degree of fit between model and data constitutes an acceptable level of reproducibility or scalability, and on how best to index this. Despite Guttman's strictures to the contrary, most published 'Guttman' scales are based on a small number of items 'purified' from a larger set. It is then possible to meet 'acceptable' levels of reproducibility (coefficient of reproducibility, Rep = 0.90) at what may in fact be very low levels of internal consistency. Even completely independent items can give a Rep of 0.90 when they are few in number and have widely spaced marginals. Various improved indices which assess degrees of scalability against a bench-mark of zero have been proposed (for discussion and further references see Moser and Kalton (1971), and

Schooler, (1969)). Following this line, however, it becomes harder to establish reasons for preferring a cumulative scale approach to other internal consistency methods, of which the Likert procedure is a main instance. When the hierarchical positioning of the items has meaning in terms of an underlying phenomenon (see McKennell, 1969, pp. 4–6) then there can be a gain in information on the cumulative approach, but in the more common case we are interested in scaling the informants rather than the items. And where the departure from perfect cumulative interdependence between items is considerable, as it usually is, then measures of degrees of homogeneity calculated by either approach come to much the same thing.

Guttman scaling has proved popular with many workers because it appears to offer a rigorous basis for scale construction using only a small number of items. At the same time, more than other methods, it has been criticized because of the onus placed on the subjective judgement of the investigator in the initial selection of these items. In many attitude areas it is not easy to frame natural sounding items that have the cumulative properties demanded by the model, nor is there any compelling reason why attitude statements should always or even usually be expected to have these properties.

At a number of points in the chapter Guttman scaling is discussed again in its relation to Likert scaling since both techniques depend, in their different ways, on the analysis of internal consistency in responses to an item set. In Likert scaling however no attempt is made to assign the items to positions along the attitude continuum. In fact good Likert items will tend to fall towards either end of the continuum. The ideal Likert item should be an emphatic statement of opinion of the kind a person with strong convictions would be likely to voice in discussion on the issue. This fact makes items suitable for Likert scaling easier to devise. It is an important feature too if it is accepted, as suggested below, that the best source of items are the statements typical respondents make when expressing themselves naturally on the topic. In applying a Likert scale item respondents are asked to state the degree to which they endorse it, usually from *strongly agree* to *strongly disagree* on a five-point scale. The relation of any item to the presumed common factor is established by some kind of correlational procedure. Traditionally the method adopted is to correlate each item score with the total score summed over the set of items. Items which show a sufficiently high correlation with the total score are retained in the scale and items which do not are rejected. The final scale is then constructed to comprise only the remaining items. Because of the ease with which suitable items can be devised, and its rapidity as a method of item analysis Lemon (1973) in a recent comparison of unidimensional scaling models described the Likert technique as 'justifiably the most popular method of attitude scale construction in use today', and with this verdict we agree. Its main drawback, that it can yield an apparently unidimensional scale which is actually multidimensional, will be discussed after covering the more general alpha approach which, among other things, is designed to obviate this possibility.

Likert scales are sometimes referred to as *summative* scales since they are scored by adding the response scores on the component items. Adding the item scores together implies that each item is a linear (or at least a monotonic) function of the same attribute. The model is a probabilistic one in that it is assumed that each item is an imperfect measure of the critical attribute. This is in line with psychometric theory which postulates that the total score on a multi-item test comprises a 'true' attribute component and an uncorrelated, composite error component. Test reliability in psychometric theory refers to the proportion of total score variance attributable to the true score variance. This chapter does not dwell on the mathematical elaboration of the test theory model such as can be found in text books on the subject. It focuses rather on those aspects not sufficiently emphasized in many accounts of Likert scaling but which in the writer's experience render it a specially valuable method for the practical business of scale development. These features, of particular utility in exploratory attitude surveys, are the availability of factor and cluster analysis procedures for finding the main dimensions underlying an item set, the provision of a test of discriminant validity in terms of the reliability of measurement, applicability to attitudes at different levels of generality–specificity, a rationale for relating reliability to homogeneity and test length, and hence for trading off scale length against reliability. Considerations of these features make up the bulk of the chapter. The relevance of the issues singled out for discussion, however, will be appreciated better if we first stop briefly to consider the place the scale development phase as a whole occupies within the context of an exploratory attitude survey.

7.3 STAGES IN AN EXPLORATORY ATTITUDE SURVEY

In order to be sure of adequately exploring, representing and interpreting the structure of attitudes in the population under investigation the designer of a large scale sample survey should be prepared to carry out fieldwork in three basic stages. The characteristics of these stages are set out in Table 7.1 below. The way each stage fits into the total flow of the investigation is illustrated in Figure 7.1. For simplicity we have not included as a separate stage the conventional piloting of the main survey questionnaire that should always be done as a matter of routine. Also, we have concentrated on spelling out the three basic stages without elaborating on the many possible variants. For example, one may recycle from stage 2 to stage 1 in order to build up further the item pool if the scale development work suggests this is necessary. Similarly, limitations of existing but inadequate attitude scales may be transcended by using them to locate subjects with disparate positions as a starting point for the qualitative pilot (see McKennell, 1974b, pp. 71–72).

Most of this chapter deals with the intermediate scale development stage, but we are now concerned with setting this in perspective against the stages that come before and after. In this connection it is important that the example to be used below to illustrate details of the psychometric work at the scale development stage should be seen in the perspective of the total study within which it formed only a small part. The example will show how a small set of items brought together to tap a general attitude of approval/disapproval of smoking turned out to cover two components that differed in ways important for the study. This distinction, it is

Table 7.1. Stages in a large-scale exploratory attitude survey

	Stage	Sample size	Questionnaire characteristics	Objectives
1.	Qualitative pilot	10–50 individual interviews, 3–10 group interviews	No standardized schedule is used	Exploration of the conceptual arena, hypothesis derivation, item elicitation
2.	Scale development pilot	100–250 interviews	Large number of standardized statements	Scale construction
3.	Main survey	1000 interviews upwards	Reduced attitude measures plus background and behavioural characteristics	Hypothesis testing; construct validity checks; search for attitude correlates

believed, may well have been missed if the items chosen for scaling in the first place had not been reflections of statements made by typical respondents at the qualitative stage. The study of smoking habits and attitudes (McKennell and Thomas, 1967) from which the example was taken began with an initial qualitative stage consisting of eight group discussions and twenty extensive interviews with smokers and ex-smokers mixed by age, sex and social class. It was followed by a scale development pilot on 125 adult and adolescent smokers and ex-smokers. The sample for the main survey stage consisted of 854 adolescents and 984 adults. In addition to the two dimensions to be discussed in the example the qualitative work generated items for many other attitudinal variables related to smoking. These were

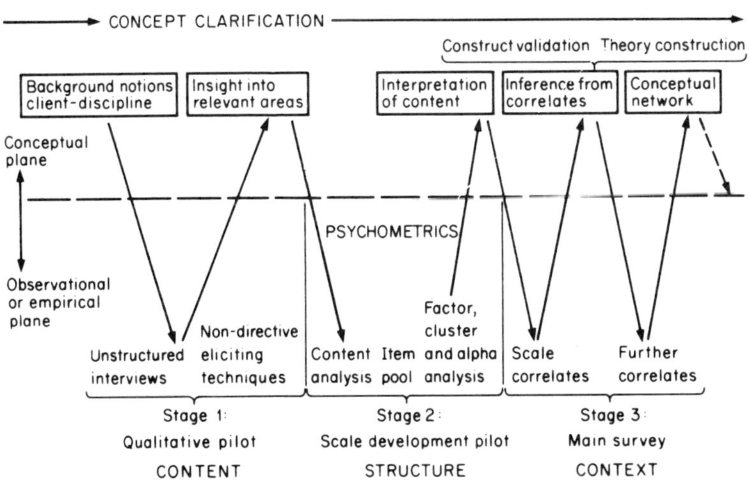

Figure 7.1. Attitude measurement in exploratory surveys. (Reproduced with permission from A. C. McKennell, *Surveying Attitude Structures*, pp. 45–55, Elsevier, Amsterdam, 1974).

further distilled at the scale development stage so that the final schedule contained reduced attitudinal measures (in addition to the two to be considered in the example) relating to the following: degree of belief in the lung-cancer/smoking connection, belief in the connection between smoking and other ailments, belief in the extent that one was personally at risk in these respects, attitudes to the expense of smoking, to the nuisance and aesthetic aspects of smoking, to facts of the Government's anti-smoking campaign that were found to scale together and to another facet, the banning of smoking in public places which (importantly) was found not to belong on the same scale. Hypotheses generated at the qualitative stage were tested and explored further during the analysis of the main survey data. At the main survey stage the refined measures of the attitudinal variables distilled from the scale development stage were related to each other and to behavioural variables in a total explanatory scheme.

Before proceeding with the main discussion on psychometric techniques further comments are in order on the initial qualitative exploratory phase. Discussion of the final, main survey stage will be found in Section 7.5.

7.3.1 Stage 1: The Qualitative Pilot

Once it is admitted that more than one item is required, the problem arises of how to go about generating a set of items to represent an attitude domain. The term 'universe of content' has become current in the literature to cover the notion that the items actually used can be regarded as a sample from all the statements that could conceivably be produced in reference to the attitude domain under consideration. The term 'universe' has, however, become bound up with attempts to systematize the sampling analogy by way of *a priori* definitions on the part of the investigator. We are not interested, moreover, in sampling from anything and everything that can be said about the attitude object, but only from those aspects that are salient in determining the evaluations placed upon it. Salience here refers to what is phenomenally relevant in the outlook of the respondents whose attitudes are being investigated. We have therefore used the alternative term 'conceptual arena' in order to stress the procedural implications of the fact that the attitudes being investigated are after all located in the minds of our respondents. The given issue or attitude object, the required domain of attitude responses, may be circumscribed broadly by the investigator's initial conception of what is relevant for his research purposes, but its elements can only be tentatively fixed in advance. The elements sought are those that are salient in the common conceptual arena of the population under study.

This standpoint, which has been elaborated more extensively elsewhere (McKennell, 1974), implies that initial recourse must be made to representative informants to discover what elements are salient for them. To put it at its simplest, in order to understand how people think it is first necessary to make arrangements to hear what they have to say. Such a conclusion might seem redundant, and possibly a bit pious, were it not for the constancy with which the preliminary stage of qualitative exploration is ignored or skimped in much attitude research.

Schedules used to generate quantitative data have necessarily to be standardized around their designer's preconceptions. Preliminary qualitative work is therefore essential if the investigator is to be safeguarded from his initial biases and if the item battery finally used is to reflect what is important in the outlook of informants themselves. The pilot work at the initial item-generating phase should therefore utilize non-directive approaches which do not rely on a fixed schedule of questions. There is a dearth of good writing on procedures for the preliminary stage but the booklet on Depth Interviews and Group Discussions by Social Community Planning Research (1972) can be recommended.(The article by Frost and Braine (1967) is worth consulting for the use of 'repertory grid' devices for item eliciting.) Essentially what is required is a series of free, extended interviews and group discussions, without a formal questionnaire, with a small sample of respondents typical of the population to be surveyed. The essence of the non-directive approach is that it is non-leading: the investigator must be scrupulous to proceed in such a fashion that does not get a playback of his own attitudes or what he regards as important. This phase of the work therefore provides the maximum opportunity for respondents to reveal the reasons underlying their evaluations, for saying what in their view is related to what in the attitude domain and for introducing elements that stem naturally from their attitudes. Such interviews, preferably recorded on tape, serve two purposes: they provide a rich source of hypotheses (to be tested at the main survey stage) concerning the origins, complexity and consequences of the attitudes in question, and they yield vivid expressions of opinion and points of view in the natural language of the population. These expressions, turned into a standardized form which preserves the flavour of the original, are the best source of the questionnaire statements or items for the scale development stage. The item pool arrived at by this procedure may be very large, but the quest for representativeness requires that any error should be on the side of comprehensiveness. It will be part of the business of the subsequent scale development stage to reduce the items to a manageable number. No scaling model nor any kind of subsequent statistical analysis, however sophisticated, can make up for omissions at this initial qualitative stage.

The survey on smoking habits and attitudes noted earlier is typical in that it shows how qualitative exploratory work will reveal the complexities and ramifications in what may have initially seemed a simple attitudinal area. It is the essence of good attitude research to be prepared to explore and measure as a complex topic area what may superficially seem to be a single attitudinal object. At the same time we do not believe that justice can be done to the representation of an attitude structure by moving immediately from the results of a qualitative pilot to the design of a main survey schedule, as Lazarsfeld (1944) once suggested. For how is one to cope with the problem of isolating one attitude from a considerable number of others which in their turn may be part of a wider attitude system, or with the lengthy questionnaires that seem to be required if so many attitudes are to be probed and each with a multi-item scale? It is the business of the intervening scale development stage, discussed next, to provide techniques for deciding how many dimensions should be measured, for testing whether one attitudinal variable is really distinct from another, for measurement at different levels in the same domain, and for reducing the number of items required to measure any one variable.

7.3.2 Stage 2: Scale Development Pilot

Limitations of Unidimensional Approaches

Granted that a large item pool has been assembled and that it contains indicators of the salient elements in the common concept arena, the problem is that of establishing scales to measure evaluative positions along each element. With only unidimensional scaling procedures available, the first step would need to be the demarcation of the total item pool into subsets of items. Unidimensional scaling procedures would then be mounted for each subset. The snag about proceeding on an *a priori* basis to specify more refined subdomains is that the variety of subclassifications that are possible becomes almost endless. Different investigators may want to impose different classifications and reshuffle the items accordingly. A large variety of alternative, 'logical' classifications could be made. Many of the rearrangements would appear reasonable both to the investigator and independent 'judges' since the human mind has a great capacity for reading meaning into even haphazard arrangements of verbal materials.

Obviously the demarcations that are needed are those that are important from the informants' perspective. If what has been sundered in the investigator's mind actually belongs together in the conceptual arena under investigation, then certain limited objective checks are available on the basis of the informants' responses to the items. Briefly, as we discuss further below, the intercorrelations of the separate scales can be obtained and compared with their reliabilities. The use of two scales rather than one would be justified only if the scales intercorrelate at a level lower than would be expected from a knowledge of their reliabilities.

The real difficulties begin when the investigator has unwittingly included two or more scalable dimensions in one subset of items. In the extreme case, if his intuitive judgement is badly astray, the subset may include dimensions that are largely independent. Even using Likert scaling, which is quite generous on the level of internal consistency required for scale construction, he might then be led to conclude erroneously that no organizing principle underlay the informants' responses to the items. More likely he would sample two or more separate but overlapping dimensions. On the strict application of Guttman scaling he would still conclude that the domain was unscalable. Using either the Likert approach, or the lower levels of consistency (reproducibility) that are usually accepted, often inadvertently, in Guttman scaling or(strictly) cumulative scaling, then a scale could be constructed. What it would measure would be the general factor in the item subset.

Here we have the main drawback in using unidimensional scaling approaches to the mapping of attitude structure. Distinctions of importance to substantial numbers of informants may well be missed in the process of imposing a hypothesis of undimensionality on the data. In the common case of overlapping dimensions the hypothesis is not so much false as over-simple. It fails to do justice to the true complexity of the attitude structure. The process of exploring the common conceptual arena which began at the qualitative stage would thus be blunted at the quantification stage. We are again obliged to invest heavily in the investigator's initial preconceptions. The possibilities of circularity, of getting out only what is put in, remain too great for complacency. What we are seeking to engage with and quantify are the 'psycho-logical' relations and distinctions in the thinking of the informants, not just those in the logic of the investigator.

Multidimensional Approaches

What is required is an extension to the quantitative stage of the intention behind the preliminary qualitative phase of attitude research, namely to safeguard the investigator from building into the measuring instrument at too early a stage classifications which stem from his preconceptions rather than from the population under study. The data, in effect, should be permitted to 'speak for themselves' with regard to the number and nature of the dimensions that are present.

Out of the variety of techniques for finding the main dimensions underlying an item set by far the simplest is a cluster analysis (see Chapter 3) in which items are assigned to subsets so that the level of intercorrelation of items within a subset is high and that between subsets is low. The same objective is implied in constructing homogeneous tests by means of factor analysis (see Chapter 4). The possibility that one of the newer, more formally acceptable models for multidimensional scaling might serve the same purpose is discussed towards the end of the chapter. We favour a cluster and, up to a point, a factor analysis approach in the present context primarily because it is consistent with and indeed may be considered part of the psychometric test tradition. That tradition may leave much to be desired in respect of its metric and other assumptions but it provides several features, centred round the concept of reliability, for which there is as yet no systematically developed alternative. These features, as listed earlier and developed throughout the chapter, are what render it especially suitable for the scale development stage in an attitude survey.

Classical psychometric theory provides a rationale for the Likert or summative scaling approach since this can be regarded as a single common factor model for attitude scale construction. What can be regarded as an extension of the Likert approach provides a simple route for uncovering multidimensional attitude structures. Items are judged as scalable on the Likert approach by the extent to which they 'hang together', measured by item-total correlations (r_{it}). The same assessment can be made more stringently by an examination of the intercorrelation between pairs of items (r_{ij}). As applied to the unidimensional case, the average value of these correlations ($\bar{r}_{ij}$ and $\bar{r}_{it}$) are directly related and provide alternative indices of the homogeneity of the item set. Scott (1960) shows that test homogeneity can be measured in terms of the average interitem correlation and that

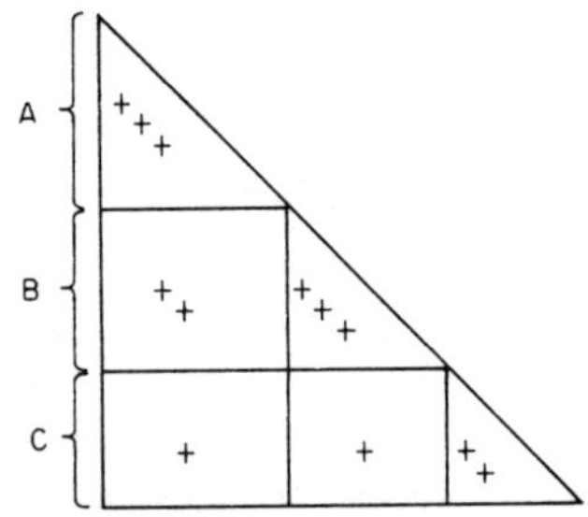

Figure 7.2. Schema of an interitem correlation matrix sectioned to show three subcomponents overlapping as described; the number of plus signs indicates the size of the correlations in a given section. (Reproduced with permission from A. C. McKennell, *Surveying Attitude Structures*, pp. 45–55, Elsevier, Amsterdam, 1974).

this is mathematically equivalent to measuring the average correlation between the item and total score and also to Loevinger's (1948) coefficient.

Nowadays using computers it is just as easy to obtain the complete matrix of $m(m-1)/2$ interitem correlations as it was formerly to obtain the m item-total correlations for a set of m items. The advantage of obtaining the complete interitem correlation matrix is that the internal structure of the item battery can be examined for the presence of major subcomponents. These will be revealed as subclusters of items which intercorrelate more highly with each other than with the remaining items.

Figure 7.2 shows schematically an interitem correlation matrix with three distinct components. The items which intercorrelate most highly form triangular submatrices along the main diagonal of the larger matrix. The off-diagonal boxes then contain the correlations at a lower level between the items in different clusters. Factor and cluster analysis procedures for obtaining this arrangement are discussed below. Once a correlation matrix has been clustered and sectioned in this way, it presents a meaningfully organized and readily inspectable set of data replacing what may have initially appeared as a chaotic mass of numbers. The items which load most highly on a factor will usually be the defining items of a cluster. An informant's score summed over these items is the measure of his position on the attitude component. Of course, few items are found which are factorially 'pure'. Moreover, even in a matrix amenable to sectioning, the cluster membership of some items will be equivocal. This makes little practical difference to the cluster scoring procedure. As Cronbach (1951) pointed out in his classical article, when each factor other than the first is present only in a few items, the general factor quickly comes to account for the dominant portion of the test variance as the test is lengthened by the addition of successive items.

The factor analytic model is designed to handle more complex situations than this. It will cope with matrices that are not amenable to sectioning into subclusters

One's dimensional interpretation will then turn on the niceties of the mathematical factor model, a position we would sooner avoid. The writer has examined many matrices in which the items have been derived from preliminary non-directive exploratory interviews and has yet to find any from which the subcomponents could not be taken out by the simpler clustering procedure. The reason would seem to be that in item collations derived from the content analysis of an adequate set of qualitative interview records, one tends to obtain sets of items which are serving as pluralistic indicators of distinct points of view. Evaluative positions corresponding to these distinct points of view are of course, just what we are seeking to measure.

7.4 THE ALPHA COEFFICIENT: A MULTIDIMENSIONAL APPROACH

7.4.1 Reliability, Homogeneity and Scale Length

The values of $\bar{r}_{ij}$ are an indication of homogeneity closely related to the coefficient of reproducibility or scalability in Guttman or cumulative scaling. The homogeneity of a scale will tend to decrease as the attitude we are trying to measure becomes more general or abstract. One virtue of the psychometric or reliability approach is that it provides a rationale for taking into account different levels of homogeneity in scale construction. While the concept of homogeneity is not sharply defined save in the formulae used to evaluate it, the key principle is that the homogeneity of a test is independent of its length (number of items). As Cronbach (1951) put it 'a gallon of homogenized milk is no more homogeneous than a quart'. The reliability of a test, on the other hand, increases as a function of its length. The basic idea is that random errors on particular items tend to cancel out so that the total score obtained on successive items tends increasingly to be determined by the common factor running through the items. Thus a set of items need not approach a perfect scale, in the Guttman sense, in order to provide reliable measurements along a single, interpretable dimension. Interpretability demands only that some degree of homogeneity or internal consistency be present in the item set. The alpha coefficient is an index of the internal consistency reliability that takes into account both homogeneity and test length. An approximation to the value of alpha can be quickly obtained for a given number of items from a knowledge of only their average intercorrelations. The degree of homogeneity may be quite small and yet the reliability of measurement can still be high, if we use enough items.

We now discuss further some of the properties of the alpha coefficient, the way its computation can easily be combined with cluster or factor analysis and how it can be adapted to provide a method of item selection for scale construction.

7.4.2 Alpha as a Coefficient of Reliability

'Alpha' is the label given by Cronbach (1951) to a particular type of coefficient which measures the reliability of a test, or item battery, in the special sense of its internal consistency. This coefficient is at the centre of modern psychometric thinking on the subject of reliability, of which only a brief and necessarily oversimplified account is given here. The main reference here is still Cronbach's classical (1951) paper. The article by Tryon (1957) has been influential in the account it gives of the relationship between various formulae.

A test is perfectly reliable if individuals obtain exactly the same score on it on two separate occasions. The degree of reliability can, in principle, be measured by testing a sample of individuals twice and finding the correlation between the two sets of scores. This correlation (or actually its square) is then the coefficient of reliability, varying between zero and unity. The test-retest method of estimating reliability is seldom practical in the survey context since it requires a second approach to the same informants. Moreover it has theoretical difficulties. The informants may remember their previous replies giving a coefficient that is spuriously high, or if the time between testings is too great there may have been a genuine change in the function being measured, resulting in a coefficient that is spuriously low.

For these reasons a split-half method of estimating reliability is often used. The series of items constituting the test is divided into two halves, and treated as two tests. The correlation between the scores on the two half-tests is then found and corrected for the effects of halving. The Spearman-Brown formula for estimating the reliability of the full length test from the correlation of halves is $2r_m/(1 + r_m)$ where r_m is the correlation between halves.

Alpha is directly related to this split-half method of estimating reliability. Several split-half coefficients can be obtained from the same testing, and these will vary somewhat, depending on which items are allocated to one or the other half-tests. It can be shown (Cronbach (1951)) that alpha is the mean of all possible split-half coefficients.

The case for knowing the reliability of measuring instruments should be self-evident. A variable measured with zero or near zero reliability would show no correlation with any other variable whatsoever. The researcher might then conclude that it was unimportant. In fact, it could be a very important variable, and the correct conclusion would be that he had failed to measure it. In general, the maximum correlation any variable can have with another is limited by the reliability with which it is measured. Two variables may be perfectly correlated in reality, but the obtained correlation will be low if the reliability of the measuring instruments is low. From a knowledge of the reliability coefficient, however, it is possible to calculate what the true correlation would be if one had a perfect measuring instrument. The true correlation between variables x and y is $r_{xy}/\sqrt{(r_{xx}r_{yy})}$ known as the 'correction for attenuation' where r_{xy} is the obtained correlation, and r_{xx} and r_{yy} are the reliability coefficients for x and y.

These are important uses of the reliability coefficient which make a knowledge of the alpha value for a test important in its own right. In this chapter, however, we are concerned less with these applications than with the way a special version of the formula for alpha can be adapted for test construction purposes along with factor or cluster analysis.

7.4.3 The 'Short-Cut' Formula for Coefficient Alpha

The alpha test construction procedure is based on the properties of the following formula:

$$\text{Alpha} = \frac{m\bar{r}_{ij}}{1 + (m-1)\bar{r}_{ij}} \qquad (1)$$

where m = the number of separate items in the test; $\bar{r}_{ij}$ = the average of all the interitem correlations; and alpha is the reliability of the total score obtained by summing the score on the separate items. This form of the formula for coefficient alpha is not always given in the textbooks. An easy way of seeing the derivation of this form of the alpha coefficient is by way of the Spearman-Brown formula mentioned above. For the general case this is

$$r_{mm} = \frac{mr_{ii}}{1 + (m-1)r_{ii}}$$

where r_{ii} is the reliability of the test of unit length and r_{mm} is the reliability of a test m times as long.

If r_{ii} is set equal to r_{ij} (i.e. the reliability of an element or item in the test) then r_{mm} is the reliability of the total test. This is the form of the alpha formula used here. The derivation assumes all items have equal reliabilities, equal intercorrelations and equal frequency of endorsement (i.e. equal variance). These assumptions are involved in the Spearman-Brown formula. The necessity of these theoretical assumptions has been questioned by Tryon (1957) amongst others.

Formula (1) is an approximation for alpha. A more rigorous application of reliability theory takes into account other features of the data including particularly the variance of the items. Bohrnstedt's (1969) 'quick' method yields this rather more rigorous determination. Computation by $\bar{r}_{ij}$, however, is much simpler, and in the present writer's experience, leads to closely similar values in practice. The index of homogeneity is not strictly $\bar{r}_{ij}$ but the coefficient obtained by working backwards from the more accurate determination of the alpha coefficient. Strictly it is the intercorrelation which would produce the alpha coefficient if all items had equal intercorrelations and variances. Again the discrepancy between the quick estimate and the exact theoretical value is not usually large enough to be of practical importance. For a set of items not capable of division into discrete item clusters, the value of alpha is a very good estimate of the proportion of variance due to the first principal component running through the item set. Otherwise alpha is an upper bound for this estimate. The approximation to the correct value will be very close, however, under the conditions of application recommended in this chapter, namely when applied to the items in a relatively homogeneous cluster. The virtue of using this particular formula is the short-cut it provides to the estimation of alpha whenever the intercorrelations between items are available, as they are in cluster or factor analysis. It is then a simple matter to

Table 7.2. Alpha values by approximate formula[a]

$\bar{r}_{ij}$ Average inter-item correlation	Number of items (m) 2	3	4	5	7	10	15	20	25	30	40	50
0.05	0.10	0.14	0.17	0.21	0.27	0.34	0.44	0.51	0.57	0.61	0.68	0.72
0.10	0.18	0.25	0.31	0.36	0.44	0.53	0.63	0.69	0.74	0.77	0.82	0.85
0.15	0.26	0.35	0.41	0.47	0.55	0.64	0.73	0.78	0.81	0.84	0.88	0.90
0.20	0.33	0.43	0.50	0.56	0.64	0.71	0.79	0.83	0.86	0.88	0.91	0.93
0.25	0.40	0.50	0.57	0.63	0.70	0.77	0.83	0.87	0.89	0.91	0.93	0.94
0.30	0.46	0.56	0.63	0.68	0.75	0.81	0.87	0.90	0.91	0.92	0.94	0.96
0.35	0.52	0.62	0.68	0.73	0.79	0.84	0.89	0.92	0.93	0.94	0.96	0.96
0.40	0.57	0.67	0.73	0.77	0.82	0.87	0.91	0.93	0.94	0.95	0.96	0.97
0.45	0.62	0.71	0.77	0.80	0.85	0.89	0.92	0.94	0.95	0.96	0.97	0.98
0.50	0.67	0.75	0.80	0.83	0.88	0.91	0.94	0.95	0.96	0.97	0.98	0.98
0.55	0.71	0.79	0.83	0.86	0.90	0.92	0.95	0.96	0.97	0.97	0.98	0.98
0.60	0.75	0.82	0.86	0.88	0.91	0.94	0.96	0.97	0.97	0.98	0.98	0.99
0.65	0.79	0.85	0.88	0.90	0.93	0.95	0.97	0.97	0.98	0.98	0.99	0.99
0.70	0.82	0.88	0.90	0.92	0.94	0.96	0.97	0.98	0.98	0.99	0.99	0.99
0.75	0.86	0.90	0.92	0.94	0.95	0.97	0.98	0.98	0.99	0.99	0.99	0.99
0.80	0.89	0.92	0.94	0.95	0.97	0.98	0.98	0.98	0.99	0.99	0.99	0.100

[a]Approximate formula: $\dfrac{m\bar{r}_{ij}}{1 + (m - 1)\,\bar{r}_{ij}}$

take any subset of items and compute the average of all their intercorrelations. This value, $\bar{r}_{ij}$, may then be entered into the above formula together with the value of m, the number of items. The value of alpha, so obtained, is an estimate of the reliability of the m items used as a test battery. It is not necessary to solve the formula directly each time, if use is made of Table 7.2. This gives the values for alpha corresponding to different combinations of m and $\bar{r}_{ij}$.

Another virtue of the above formula is that it brings out very well how reliability of a test depends on its internal consistency or homogeneity ($\bar{r}_{ij}$) and its length (m). This can be seen on common sense grounds. With perfectly reliable items all measuring the same thing, any one of the items would be all that was necessary to measure the particular dimension with perfect reliability. For such items the value of $\bar{r}_{ij}$, and the value of alpha, would be unity. In practice, however, when an informant responds to a single item his reply may be haphazard or, what comes to the same thing statistically, may be determined by some idiosyncratic interpretation of the item. If a proportion of informants share the same idiosyncratic point of view or if several factors are important in the response, we have a multifactorial situation. We have left this out of account in order to simplify the essence of the argument. As a result of these response 'errors' even items which appear from their content to tap the same dimension hardly ever intercorrelate perfectly. Values of $\bar{r}_{ij}$ vary widely with the topic area under investigation, but seldom exceed 0.5. Unreliability at the item level is of course the main justification for scaling. Since the errors on particular items are random they tend to cancel out, and the total score obtained by adding the scores on successive items tends increasingly to be determined by the common factor running through the items.

Cronbach (1951) shows in fact that alpha will be an estimate of the first factor variance in a test unless it contains distinct clusters of items. A test will not of course contain distinct clusters if it consists of items which themselves define a single cluster or factor, as recommended in this chapter.

7.4.4 Use of Alpha with Factor and Cluster Analysis

The use of alpha in attitude scale construction will be illustrated with some data selected from a survey on smoking habits and attitudes (McKennell and Thomas, 1967). Table 7.3 shows the matrix of intercorrelations between twelve statements of opinion about smoking. Opinions concerning the health effects of smoking were studied in another matrix leading to a separate scale (actually scales) concerned with the health aspect. The objective in Table 7.3 was to construct a scale to measure in general terms, without reference to specific reasons, the degree of approval or disapproval of smoking. On analysis, the correlations between the items in this putative scale revealed two distinct clusters, as reflected in Table 7.3.

Table 7.3 is a clustered correlation matrix. That is, the original correlations among the items have been rearranged so that the correlation matrix itself reveals the clusters and the relation between clusters. This is achieved by renumbering the items in the study so those that cluster together are in sequence along the borders of the correlation matrix. This has the effect of throwing on to the main diagonal of the matrix the triangular submatrices of relatively high correlations corresponding to each cluster. The off-diagonal boxes then contain the correlations at a lower level between the items in different clusters. Only a two-cluster matrix is being considered for simplicity, but the principle is the same when several distinct clusters are found in the initial matrix. The relations between c clusters will be shown in $c(c-1)/2$ off-diagonal boxes in the clustered correlation matrix.

Before this useful arrangement can be secured it is necessary to specify some assignment rule for allocating items to clusters. McQuitty's (1957) Elementary Linkage Analysis can be used to locate clusters of items even in very large matrices in a matter of minutes. The method is only approximate however since its speed depends on assigning each item to the cluster with which it has the highest single correlation. This will usually but not always be the cluster with which it has the highest average correlation. Item 7 in Table 7.3, for example, belongs to the first cluster on the latter assignment rule, but would be placed in the second on the McQuitty rule. Occasional large discrepancies can arise between the results of the two assignment rules, especially where the largest single correlation for an item is a chance result of sampling variability. Assignment on the basis of the highest average correlation on the other hand will tend to optimize $\bar{r}_{ij}$ and hence alpha for the cluster, and is therefore to be preferred on reliability theory.

An exact clustering on the latter basis can be achieved using Holtzinger and Harman's (Harman, 1967) B-coefficient but the method is laborious. The same result can be achieved much more quickly by clustering and sectioning a matrix using the McQuitty method as a first approximation. Any gross errors of assignment are detectable by the appearance of runs of high correlations in the rows or

Table 7.3. Intercorrelations[a] between 12 opinion statements on smoking

Smoking costs more than the pleasure's worth	x											
I hope(d) my own children never smoke (d)	27	x										
Children should be discouraged from smoking	23	39	x									
Smoking is a nuisance	33	40	33	x	Cluster A							
Smoking is like drug taking	37	36	26	50	x							
People who don't smoke shouldn't start	29	43	36	34	45	x						
Smoking does more good than harm (disagree)	25	29	19	31	25	26	x					
Young people smoke to feel grown up	12	15	16	24	21	14	23	x		Cluster B		
Smoking is a dirty habit	24	35	21	46	44	36	27	18	x			
Smoking is pleasurable (disagree)	28	21	16	35	25	13	33	03	30	x		
Smoking can help people relax (disagree)	22	15	09	24	14	13	24	05	22	50	x	
Smoking can help nervousness (disagree)	16	09	08	18	11	05	16	00	20	37	64	x
Item number	1	2	3	4	5	6	7	8	9	10	11	12

[a]Decimal point omitted.

columns of an off-diagonal box. An offending item can then be reallocated to the cluster with which it has the highest average correlation. Other methods of cluster analysis are described in Chapter 3.

If a computer is used to obtain the correlation matrix, and the relevant program is available, it can be as easy as not to factor the matrix in the same computer run. The results of the factor analysis can then be used to cluster the correlation matrix. For example, Table 7.4 shows the Varimax factors obtained from rotating the first two principal components extracted from Table 7.3. The important thing to note is that although the items load on more than one factor, the items which have the highest loading on a Varimax factor are also the defining items for a cluster. By assigning each item to the factor on which it has the highest loading, therefore, we can achieve a cluster analysis with the desired property of maximum $\bar{r}_{ij}$ values within clusters. However, this neat result cannot always be relied upon to occur even for matrices that show pronounced clusters. For example, when three principal components were extracted from Table 7.3 and rotated, it was found that the first cluster was broken up, with items 7 and 8 falling on a separate factor. When four components were rotated, the remaining items in cluster A split further, with items 2, 3 and 6 falling on one factor and items 1, 4, 5 and 9 on another. Since there is no way of knowing how many components to rotate in advance, the results of a factor analysis will sometimes disagree with a cluster analysis of the same matrix. Nevertheless in using factor analysis to construct homogeneous tests it is desirable to have factors that correspond to clusters. That is to say, the items of a group

Table 7.4. Factor loadings, $\bar{r}_{ij}$* and r_{it} values[a] for the items in Table 7.3. (Reproduced with permission from A. C. McKennell, *Surveying Attitude Structures*, pp. 45–55, Elsevier, Amsterdam, 1974)

	Principal components		Varimax factor loadings		$\bar{r}_{ij}$*		
Item no.	(I)	(II)	(A)	(B)	(A)	(B)	r_{it}
1.	56	02	49	27	26	22	55
2.	62	27	68	08	33	15	60
3.	50	28	58	01	27	11	45
4.	72	11	68	27	36	26	69
5.	67	26	71	12	40	17	63
6.	60	35	70	00	33	10	53
7.	55	03	45	31	26	24	54
8.	32	25	40	05	18	03	27
9.	65	07	59	26	31	24	64
10.	57	48	24	70	22	57	60
11.	50	70	06	86	16	42	52
12.	40	71	02	82	11	49	43

[a]Decimal points omitted.

$\bar{r}_{ij}$* Average of the correlations of each item with items in cluster A or cluster B.

r_{it} Correlation of each item with the total score summated over the twelve items.

identifying a factor should have substantially higher intercorrelations with each other than with the remaining items in the total set. In this sense, and for this purpose, cluster analysis is not only the simpler technique, it is more basic. Factor analysis of the matrix is not necessary, but if factors are obtained they can serve as the starting point for a cluster analysis which in turn can serve as a check on the factor analysis. Where the cluster solution obtained from factoring turns out to be only approximate it can be corrected in the same way as one obtained from a McQuitty analysis, as described earlier.

Having completed the cluster analysis the next step is to consider the properties of the scales which can be constructed from each subset of items. The method of clustering will have ensured that the maximum attainable homogeneity has been achieved. The reliability of measurement can be ascertained by entering Table 7.2 with the values of m and $\bar{r}_{ij}$ to find alpha for the cluster. Thus the nine item cluster in Table 7.3 has an $\bar{r}_{ij}$ of 0.30 yielding an alpha of 0.79; the three item cluster with $\bar{r}_{ij} = 0.50$ has an alpha of 0.75. The adequacy of these alpha values and the problem of test length and item selection will be discussed further below. Table 3 however, offers a third possibility for scale construction. If item 8 is disregarded, all the remaining items are positively correlated. This is sufficient to show that running through the matrix is a general factor, which could be measured using a battery of items for both clusters. This statement is confirmed by the occurrence of a first principal component (Table 7.4) which accounts for a substantial proportion of the variance in the matrix. When a general factor is present in the matrix it is usually possible as well as desirable to choose in advance (but also permissible to fix afterwards) the direction of scoring of each item which yields consistently positive correlations in the matrix. For the items in Table 7.3, for example, responses unfavourable to smoking were consistently scored high. The $\bar{r}_{ij}$ and alpha values for the entire matrix (excluding items 8) are 0.27 and 0.79, respectively. Homogeneity is lowered by seeking to measure what is common to both clusters, but this is compensated for in the reliability of measurement by the inclusion of more items. These latter results serve to illustrate the possibility, discussed further below, of measurement at different levels in the same attitude domain.

7.4.5 Transition from Scale Development Stage to Main Survey

We have recommended that on the basis of thorough-going exploration at the qualitative interview stage, a large pool of items be assembled for scale development. The work of scale development is then so conducted, as described above, that it is an extension to the quantitative level of the exploration process. The outcome should be the discovery of the number of dimensions which it is necessary to measure in order to do justice to the fine detail of attitude structure, the salient components in the common conceptual arena. The defence in depth concerning the scalable properties of the domain will rest in large part on the findings at this scale development stage.

To the extent that the work of exploration, qualitative and quantitative, has been thorough, the item battery on which data has been obtained is likely to be larger than is essential for adequate measurement of all the components uncovered. Social surveys are rarely carried out solely to operationalize an attitudinal concept. More often scale construction is undertaken to provide more refined and reliable variables than do the answers to the individual items, but only as a step in preparation for further analysis. Short scales are therefore at a premium in the main survey. Apart from problems of rapport in the interview situation, questionnaire space and interview time must be reserved for enquiry into the demographic, sociological and behavioural variables with which the attitude data will ultimately be related. As will be discussed further below, it is only in terms of the network of relationships formed with these other variables that the construct validity of the attitude measures can be established.

In concluding the scale development stage therefore the investigator has to decide how many items, and which particular items, need to be carried forward into the main survey schedule in order to measure a particular attitudinal variable. Where scales tapping aspects of the same domain are highly correlated he may wish to carry out further checks on their distinctiveness (discriminant validity) and to consider whether it is possible to get the best of both worlds, that is, by measuring both the separate aspects and the general factor that relates them. We now consider the operational procedures that can aid these decisions.

7.4.6 Use of Alpha in Decisions on Test Length

As long as the scale construction work contains a genuine element of exploration the possibility exists of discovering clusters that are under-represented in the item sample as well as other clusters that are larger than necessary for reliable measurement. If, as is desirable, the scaling work has been done prior to the main study, decisions will be required on how many (and which) items to discard from the large clusters, and whether to derive further items to build up small clusters. There can of course be no standard of length relevant to all attitude measures since, by formula (1), the number of items necessary to reach a given reliability level varies with the homogeneity of the item battery. As noted earlier, for example, cluster A ($\bar{r}_{ij} = 0.30$) has an alpha of 0.79, cluster B ($\bar{r}_{ij} = 0.50$) an alpha of 0.75. We can measure almost as reliably with the three items as with the entire set of nine items available for the first cluster. The consequences for reliability of selecting a subset of items to represent cluster A can easily be determined from formula (1). The detailed item selection procedure is discussed in the next section. But we may note that four items (2, 4, 5 and 6) taken alone, have an $\bar{r}_{ij}$ of 0.41 and hence an alpha of 0.73. The rise in homogeneity (0.30 to 0.41) has not quite compensated for the drop from nine to four items for cluster A, but the saving in questionnaire space was here judged well worth the marginal loss in reliability (0.79 to 0.73). In the case of cluster B it was judged that although only three items were available, the level of reliability reached was not so low as to warrant a further cycle of scale development work to obtain more items.

Obviously, the length of scale deemed adequate in any situation depends in its turn on the level of reliability we are prepared to accept as adequate. What value should be aimed at? The answer must vary with the nature of the research and the purpose for which the scale is required, and in any case cannot be pursued here. For further discussion of this issue see McKennell (1970, pp. 235–238). The achievement of very high levels of reliability would seem to be a practical requirement mainly when decisions affecting specific individuals hinge on the measurement results. Such decisions are required, for example, in the use of mental tests for grading or selection purposes, but hardly ever in sample survey research. The survey analyst may have theoretical reasons for wishing to know what would be the 'true' correlation between two variables in the absence of measurement error. If so, he can use the knowledge of the reliability actually achieved to apply the correction for attenuation. More typically, inferences from survey data rest on comparisons of the relative magnitude of associations. Such inferences may hold good even though each observed association is too low due to measurement error, provided that the attentuation effects are spread more or less evenly over all the associations compared. The fact that this condition would usually hold in applications of the traditional logic of survey analysis (Lazarsfeld, 1955; Hyman, 1955; Rosenberg, 1968) may be a reason why neglect of the reliability issue has not been crucial Basically, that logic called for examining only a single association, in the form of a bivariate table, but within subgroups formed by another variable or two. Modern computer based schemes for multivariate analysis, on the other hand, typically examine a whole network of relations simultaneously, with correlations, or other coefficients of association, being the input data to which the statistical model is fitted. Distortion of the pattern of the observed relationships due to measurement error thus becomes a more serious possibility. Particular relationships may be corrected for attenuation but wholesale correction of an input correlation matrix is not to be recommended as a routine procedure. It can run into various difficulties without a careful analysis of underlying assumptions (e.g. there will be over-correction if two variables share a common factor, including a method factor). Recent extensions of path models to include the study of measurement error (Alwin, 1974; Hauser and Goldberger, 1971; McKennell, 1975), offer a promising tool for such analysis. Correction for attenuation in fact becomes implicit when reliability estimates are incorporated in a path model for 'true' or unobserved variables.

For some purposes, if not most, precise information on the extent to which one's measures fall short of the highest levels of reliability can be as useful to the sophisticated analyst as the actual attainment of such levels. But the immediate point is that in following the alpha approach at the scale development stage the researcher is not only informed about the reliability of his measures but is in a position to balance the number of scale items against the level of reliability he is prepared to accept or wants to attain.

7.4.7 Use of Alpha in Item Selection

Once it is decided to use only a subset of the items in a cluster, a technique is

Table 7.5. Reliability after discarding successive items. (Reproduced with permission from A. C. McKennell, *Surveying Attitude Structures*, pp. 45–55, Elsevier, Amsterdam, 1974)

Item No.	8	7	3	1	9	6	2	4	5
m	9	8	7	6	5	4	3	2	–
$\bar{r}_{ij}^*$	0.18	0.26	0.27	0.26	0.31	0.33	0.33	0.36	0.40
$\bar{r}_{ij}$	0.30	0.33	0.35	0.37	0.40	0.41	0.42	0.50	–
Alpha	0.79	0.79	0.79	0.76	0.77	0.73	0.69	0.67	–

m = number of items remaining in the scale after item on the left discarded.
$\bar{r}_{ij}^*$ = initial correlation of each item with the other eight items.
$\bar{r}_{ij}$ = average intercorrelation of the m items.
Alpha = reliability value for the m items.

required for selecting the best items from the pool. By formula (1), for a given number of items m, reliability will be greatest when $\bar{r}_{ij}$ is a maximum. Each item will contribute to $\bar{r}_{ij}$ according to its average correlation with the other items in the cluster. It is a simple matter to compute this value, which we shall term $\bar{r}_{ij}^*$ (to distinguish it from $\bar{r}_{ij}$, the average for all the items in the cluster). If the items are ordered in terms of $\bar{r}_{ij}^*$, this gives the order in which they should be discarded in order to preserve the maximum possible $\bar{r}_{ij}$ and alpha values for the remaining items. Strictly new values of $\bar{r}_{ij}^*$ should be calculated after each item is discarded, but it is doubtful if such an iterative procedure is worthwhile. Important discrepancies in $\bar{r}_{ij}^*$ values would tend to remain, while minor reversal would be unimportant. Table 7.5 shows the items in cluster A, in our example, ordered by their $\bar{r}_{ij}^*$ values (taken from Table 7.4). The second row of the table gives the alpha value for the remaining items after each item has been discarded. The scale constructor thus has precise knowledge of the effect on reliability of discarding successive items. The final decision on the subset of items to be carried forward for measurement purposes can be made accordingly.

As Table 7.5 illustrates, alpha values can remain high on this approach even as the scale is shortened, since convergence is towards items with high internal consistency, representing the core, as it were, of the factor being measured. The procedure is automatic enough to be programmed for a computer. But since the calculations are not onerous it is as well to do them by hand. One can then watch that the increment in homogeneity is not being brought about by a convergence towards items which are exceptionally highly intercorrelated only because they are worded almost identically. In this case the last few retained items would be tending to measure a 'wording factor', specific to them, and different from the common factor underlying the cluster. For this reason, too, items with very similar wording, should be omitted from the sample of items used in the field work. Also, since small differences in $\bar{r}_{ij}^*$ do not greatly affect the ultimate alpha value, one need not be completely inflexible at the item selection stage. Between two items, the one with a slightly smaller $\bar{r}_{ij}^*$ value might still be retained, say, if its content looked more 'interesting' to informants, and would help rapport in the interview situation.

7.4.8 Scoring and Combining Items

Once a subset of items has been selected to form a scale, an informant's scale score is obtained by simply adding the scores on the separate items. As in Likert scaling, all the items carry equal weight. It can be shown (Richardson, 1941) that with nominal weights of unity, each item carries an effective weight (in the total score variance) proportional to its $\bar{r}_{ij}{}^*$ value. This fact was allowed for in the choice of items for the final subset. While each item in the final subset could be weighted by its $\bar{r}_{ij}{}^*$ value, there is little to be gained by introducing fractional weights. Where the range of weights is small and the items positively correlated, the correlation between the composite scores obtained with any two sets of (positive) weights is very high. This has been shown theoretically by Richardson (1941) and demonstrated empirically by Likert (1932), amongst others. By the same token, where cluster-directed factors are concerned, the use of rigorously derived weights for estimating factor scores (Harman, 1967) is not worth the trouble. (Weighting becomes a problem as the number of items is small and they vary widely in their loading on the underlying common factor. The scale construction procedures stressed in this chapter are designed to avoid this situation.) Simply scoring the defining items of the factor as unity and the rest zero would yield scores which were highly correlated and of the same reliability. This has been shown both empirically and theoretically (Horn, 1965; 1969). Where factors correspond with clusters, therefore, the same method of simple, unweighted summation can be employed to obtain scale scores.

Similar considerations lead to a rejection of the notion of using standardized scores rather than raw scores for the items. Standardizing would be the equivalent of using a fractional weight for each item inversely proportional to the standard deviation of its raw scores. Of course one should take steps to equalize any gross inequality in the range of scores if, say, a two-point response scale had been used for some items and seven-point scales for others. It is neater if all the item response scales have a standard form. The conventional five-point Likert scales (with categories ranging from Strongly Agree to Strongly Disagree) will probably give as good results as any other. But there is reason for thinking that the number of scale points is a technical detail the importance of which has been overrated. The work of Komorita and Graham (1965) and more recently Matell and Jacoby (1971) indicates that when scores are obtained by summing the ratings on a battery of bipolar scales (including the Likert type), the number of points on the individual scales makes little difference to the reliability of the total score.

7.4.9 Discriminant Validity and Interchangeability of Indices

We have suggested that in seeking to map the salient components in a common conceptual arena the smallest element on which a score is obtained should be a set of items not capable of division into discrete item clusters. In using composite scores based on multiple indicators to both measure and separate the different components of an attitude structure an important conceptual tool is the notion of interchangeability of indices. If we can specify the precise sense in which two indicators are or are not interchangeable we are in a position to determine whether or not we are tapping separate components. As noted elsewhere (McKennell, 1974b, p.20), reliability theory cannot solve the problem of defining the universe of content, but it does provide an operational solution to the problem of deciding whether two subsets of items are equivalent samples from the same universe. The actual correlation between the additive scores on the two subsets may be quite low, but this does not of itself establish that they can be employed to measure two functionally discriminable components. The level of reliability with which each component has been measured sets limits to the size of the correlation with any outside variable, and typically in social science the components will be measured with reliabilities that are far from perfect. Knowledge of the reliability of each measure, however, provides a test of the 'discriminant validity' (Campbell and Fiske, 1959) of the two components. The test is a simple one. It involves ascertaining that the correlation between the two putatively separate scales does not reach the value that can be estimated from their reliabilities on the hypothesis that they are measures of the same dimension. If two measures, p and q, have the same reliability ($r_{pp} = r_{qq}$) then for discriminant validity their intercorrelation, r_{pq}, must be less than this reliability value. Otherwise the test is

$$r_{pq} < \sqrt{r_{pp} r_{qq}} \ .$$

The two components can be accepted as measures of separate variables if their intercorrelation is lower than the (geometric) mean of their reliabilities.

One advantage of having represented in the same inter-item correlation matrix all the subsets of items which might be considered as forming separate variables is that the test for discriminant validity can be quickly and easily made without ever adding items together. As noted above, coefficient alpha, the index of internal consistency reliability, can be estimated from the average item intercorrelation and the number of items in a composite. Similarly, the correlation between two composite variables can be expressed concisely by taking account of the sums of item intercorrelations both within and between the clusters. Using only item intercorrelations, computation would follow the formula for the correlation between oblique factors. Harman (1967, p.23) gives this as

$$W_{pq} / \sqrt{(W_{pp} W_{qq})} \ ,$$

where W_{pq} represents the sum of correlations between all variables in group p with all those in group q, and W_{pp} and W_{qq} represent the sum of correlations between all variables in Groups p and q, respectively. The summations include the 'self-correlation' of each variable. This is by far the simplest procedure, though technically it involves assuming that the 'self-correlation' of an item is its communality. Most practical purposes are served by taking as the communality the highest correlation formed by an item with other items in the set being considered (Harman, 1967, p.83). A more accurate procedure involves working with the entire covariance matrix, that is, item standard deviations are brought into the computations in addition to the intercorrelations.

Actually, one should rarely need to carry out the test of discriminant validity for sections of a clustered correlation matrix. The need could arise if, for example, when on examining the correlation matrix sectioned into subclusters, as in Figure 7.2 above, it is noted that two subclusters show a high degree of overlap, in that the average level of inter-item correlation in the off-diagonal box for the two clusters is not much lower than the level of interitem correlation within each cluster. This situation occurs typically when the clustering of the correlation matrix has been based on the results of a factor analysis in which too many factors have been extracted. Otherwise the discriminant validity result is already implicit in a matrix that can be sectioned into well-defined subclusters.

The need to carry out the computations to test discriminant validity arises more often when working initially with separate interitem correlation matrices. Although correlation matrices for as many as 80 or more items can be readily formed on the computer and programs exist for rearranging a matrix to aid the examination of subclusters (e.g. Borgatta and Stolz, 1963), the inspection procedure becomes onerous for the very large item pools that may be employed at the scale development stage in a thorough-going analysis of attitude structure. Also the investigator may have good reasons for tentatively demarcating certain domains or subdomains and considering these separately. For these reasons, work may proceed on a number of separate item intercorrelation matrices. Decisions can then be made on the number of variables to be abstracted from each matrix. These variables can then be scored and their intercorrelations examined in a final summary matrix. This will give a view of the relations between all the attitudinal variables being considered, and permit tests of discriminant validity as necessary. The procedure here is similar to that which would be followed in a unidimensional scaling approach, with the all-important difference that each domain has first been carefully examined for subcomponents.

In making the tests for discriminant validity, the investigator may be faced with the situation where the correlation between two scales is only slightly less than their reliabilities. The difference may be statistically significant, but the margin of separation is small, so that doubts arise whether or not there is anything to be gained by deriving two scores rather than one from the items. The soundest procedure is to give the obtained separation the benefit of the doubt and to select items from both scales for the final questionnaire. In factor analytical terminology this advice is equivalent to the recommendation to err on the side of rotating too many rather than too few factors, and to be prepared to work with oblique factors with only a small angular separation. If the separation proved non-productive on further analysis, the combined score can then be used. What may seem a small difference in the correlational data aggregated across a sample, may turn out later to conceal a major discontinuity in attitude structure for a segment of the sample, however.

Analogy with ability measurement may serve to illustrate the issue. Tests of verbal and numerical ability, for example, intercorrelate because a substantial proportion of people tend to perform either consistently well or consistently badly on both kinds of tasks. The correlation is incomplete because (errors of measurement apart) there exists a proportion of individuals who perform well on verbal material and badly or not so well on numerical tests, and *vice versa*. Translated into attitude terms, this is comparable to the existence of two components of evaluation which a proportion of the sample treats as equivalent, but between which another segment or segments are discriminating. The latter segments may include those who are ambivalent towards the attitude issue, or those for whom one aspect is salient but not the other. Alternatively, the segments whose responses are producing the degree of overlap obtained may include those who are using an underlying organizing principle or perspective that is not generally shared by the remainder of the sample. If we can separate these segments and compare the differences in the profile of their scores across the several components, we may learn a great deal about the nature of their distinctive attitude structures.

7.4.10 Measurement at Different Levels of Generality in the Same Domain

If, as usually happens, the item intercorrelation matrix shows an overall level of positive correlation, we are in a position to measure the general factor as well as the subcomponents. (Since the direction of attitude items is arbitrary, it is usually possible to transpose the sign of any substantial negative correlations so that the intercorrelations within and between item clusters all become positive.) The analogy, if we do not press too hard, is again with the field of human ability. This can be regarded as made up of distinct abilities (such as memory, reasoning, verbal, numerical and the like) which can be measured separately. Because such measures intercorrelate we can, if we choose, concentrate on the overlap, on what is most general to the field. In effect, this is what the unidimensional scaling approaches seek to do for an attitude domain. The advantage of using a multidimensional approach is that it can provide the same information and produce the same measures as in unidimensional scaling, yet at the same time examine the internal structure of the domain and provide measures on the subcomponents or lower-order factors.

These lower-order factors concern the more specific aspects of the domain. They may be related hierarchically not only to the general factor but also to each other. In Figure 7.2 for example, components A and B are shown schematically to overlap with each other to a greater extent than either does with the third component. These differences in degree of overlap are shown reflected in Figure 7.2 by differences in the average level of the item intercorrelations found in the off-diagonal boxes. Just as we combine all three components to measure what is most general to the matrix, we can combine the two that overlap most strongly to measure a component at an intermediate level of generality. Although specialized techniques exist for the hierarchical analysis of a correlation matrix (see for example, Schmid and Leiman, 1957; Hendrickson and White, 1966; and McQuitty, 1960), in a large matrix sectioned into several subclusters it is frequently possible to pick out by eye several levels of generality at which the domain can be measured.

We are proposing that this examination of the internal structure of an attitude domain should take place at the scale development stage. At the termination of this stage the decisions, considered above, have to be made about the reduced set of items that will need to be carried forward as part of the schedule for the main survey stage. In order to do justice to the fine details of attitude structure, the best policy is to carry forward into the main survey stage sufficient items to provide indicators of all the lower-order components which analysis of the item intercorrelation matrix provides some grounds for believing can be measured as separate dimensions. Then at the main survey stage we will be able not only to measure the lower-order components, but also treat as variables the more general factors as well. Although complex psychometric considerations can arise about such combinations, in many and perhaps most studies we can obtain a useful approximation to the scores on more general factors simply by adding together the scores on the subsets of items. These items should be drawn in equal numbers from each of the lower-order components which the general factor integrates. In the case of a general factor for an entire domain, this means adding scores on items from each subcluster in the matrix.

It is therefore possible to employ the same battery of items to measure an attitude domain at different levels of generality. Usually it will be sufficient to treat the multidimensional aspects of the domain as if it were made up of only two levels. The interest will be in the informants' position on the general factor, their gross evaluative dispositions within the domain, on the one hand, and the profile of their positions on the main subcomponents of evaluation that determine this disposition, on the other. It is the pattern of scores across the lower-order factors that yields the detailed information on the informant's attitude structure. Occasionally and perhaps particularly when dealing with more abstract attitude domains, it may be useful to take a more hierarchical view of this structure, and consider how components at an intermediate level of generality subsume still less inclusive components.

By way of brief examples we may refer back to the many specific attitudes and beliefs that were measured in the survey of smoking habits and attitudes, as listed on p. 200. Many, but not all of these attitudinal variables, though established as distinct components, were nevertheless found to be linked in various degrees to a common underlying factor of approval or disapproval of smoking. The fact of this linkage had important implications for the overall strategy of anti-smoking appeals. Knowledge of the extent to which an attitude was linked to the general factor had implications for handling anti-smoking appeals in that particular area, as of course did the finding that some attitudes were independent of the common factor.

7.5 THE MAIN SURVEY

The questionnaire for the third or final stage of a thorough-going attitude survey would incorporate not only the reduced attitude measures distilled from the scale development stage but also other kinds of variables concerning demographic factors, background social structure and behaviour, with which the attitudes can be related. With a main sample size of upwards of 1000 informants, the data base will be large enough to permit a thorough exploration of these relationships. Strategies for the analysis of attitudinal data in the context of other information gathered in a large-scale sample survey are not well developed in the literature. Texts such as Hyman (1955) and Rosenberg (1968) cover the essential logic, but are limited in their neglect of both the process of attitude assessment itself and of modern computing techniques. Given these techniques and a more rigorous procedure for scale construction then the possibilities for the analysis of attitudinal relationships are much enriched. For a partial account of these further possibilities in one major field of attitude research see McKennell (1974a, 1975). Here we wish only to stress how the aim of exploring, representing, measuring and interpreting a population attitude structure is a continuing process which, as Figure 7.1 above attempts to illustrate, builds on the preliminary work in stages 1 and 2 and continues into the third stage with the analysis of main survey data. The first two stages properly carried out will ensure that attitude components, important in the sense that they will be diagnostically associated with background and behavioural variables, are not overlooked. But the measurement properties that underlie the scale development stage (reliability, homogeneity and discriminant validity), while necessary, are not a sufficient condition for useful scales or for interpreting the meaning of the scales obtained. The properties of homogeneity and reliability enable us to say that a scale is measuring some variable at a given level of accuracy: discriminant validity indicates that different things are being measured. We can be sure we are measuring something but are not yet in a position to say for certain what.

It is even possible that the reliability indices derived at the scale development stage derive from sources of internal consistency that are largely if not entirely irrelevant to the purpose of the investigation. Items could cluster and exhibit high internal consistency for several reasons. They may have tapped a narrow range of content, being slight rewordings of the same statement. This was one of the early criticisms made against the high internal consistency (high reproducibility) required by Guttman scaling (Festinger, 1947). Secondly, they may have elicited some kind of common response set independent of content. There is a vast literature on response sets. For a brief overview see Scott (1968). There is a suggestion of an acquiescent response set in Table 7.3, since items which call for disagreement tend to cluster. However, the work of construct validation, described later, shows that informants were reacting to the content of the items. A third reason is that some statistical artefact may have affected the size of the correlations. Product moment correlation coefficients should be used for computing alpha (in the case of dichotomous items these will be phi coefficients, $\sqrt{(\chi^2/N)}$). This coefficient can be at its maximum value between two items only when they both have similar response distributions. There is a considerable literature on the ramifications of this peculiarity of the coefficient. Its seriousness as a source of spurious 'difficulty factors' is a matter of dispute (e.g. Dingman (1958), Borgatta (1962)). It is possible to correct alpha for the effect of disparate item distributions, but Cronbach (1951) argues that at the level of r_{ij} normally encountered, the correction makes not enough difference to be worthwhile. It is characteristic of the psychometric tradition to make strong assumptions about the data (e.g. variables normally distributed on interval scales) in order to use classical statistical theory, and then test the effects of departure from these assumptions.

Even if all these artefacts can be discounted we do not yet have sufficient basis for interpreting what the scales mean in terms of the informants' attitudes. The validity problem, in short, is wider than that of reliability. Validity concerns the extent to which some useful concept is being measured. What we have to validate is the correctness of our interpretation of the measure in terms of the concept.

A preliminary interpretation can be obtained simply by looking at the apparent meaning common to the items which cluster. We can start but should not conclude the search for interpretation in this way. The issue arises at the stage of scale development noted in Figure 7.1 by the arrow pointing to the box entitled 'Interpretation of Content'. Naming a scale at this point does not thereby confer on it all the name connotes nor make it an independently measurable entity. Far from item content being a source of validity, the interpretation of content is something which itself calls for validation. What we can get from the interpretation of content is a suggestion of where to look for the scale correlates. The next arrow in Figure 7.1 therefore points to the analysis of main survey data, where the search for correlates can go on with the additional variables and the large data base that are made available at this stage.

To illustrate briefly with the example from Table 7.3. Should the common factor in Table 7.3 be measured? Are the subfactors corresponding to the clusters worth measuring on their own? These questions cannot be answered solely by reference to homogeneity, reliability or even the discriminant validity test. The precise way in which the meaning of items 10, 11 and 12 in Table 7.3 differed from that of the remaining items was not immediately obvious, for example. Initially, some kind of artefact was suspected. Nevertheless it was decided to carry forward items from both clusters into the main survey. Analysing the main survey data it was then found that non-smokers were more negative than smokers on both of the two scales derived from clusters A and B in Figure 7.3. This is consistent with the existence of an underlying general factor. The difference between smokers and non-smokers was much greater, however, on scale B than scale A. Moreover, continuing the analysis, it was found that scale B, but not scale A, correlated with indicators of the need to smoke measured independently at another point in the questionnaire. Scale A on the other hand was linked to specific negative attitudes to smoking (belief in the health risk and favourable opinions about anti-smoking campaigns) to a much greater extent that scale B. The pattern of correlations so far supported an interpretation of scale A as measuring general evaluation (like–dislike) of smoking, and scale B as measuring belief in the extent to which smoking serves a definite function for the smoker. This interpretation was strengthened on examining the characteristics of a subgroup of smokers whose attitudes were decidely negative on scale A but positive on scale B. These turned out to be heavy, addicted smokers, who would like to give up and had made frequent unsuccessful attempts to do so. Both scales were necessary to record the fact that these smokers' attitudes were ambivalent, in that they acknowledged yet resented the hold their habit had over them. In this instance, the decision to carry forward the group factors found in the pilot and work with them in the main study was justified by the results.

In general it can be argued that the soundest procedure is to give an obtained cluster the benefit of any doubt, use the clustered items (or a selection of them) as a scale in the main survey and then base the judgement of the utility of the scale, and the interpretation of its meaning, on its pattern of correlations with other variables in the study.

The entire process of elaborating and verifying the meaning of scales, after their construction, in terms of a whole network of relationships with other variables in the survey, is akin to what Cronbach (1955) has described as 'construct validation'. (For a fuller discussion of this issue as it applies to attitude surveys see the section 'Construct Validity and the Objectivity-Projectivity Continuum' in McKennell (1974a)). It seems a sound procedure to follow when the aim is to explore and leave room for real discoveries about the structure of attitudes in the population. The investigator is protected even during the last stage of the process of scale construction, in interpreting the scales, from the consequences of his, possibly faulty, preconceptions about the number and nature of important attitude dimensions in the area under study. This information is still allowed to emerge as far as possible from the responses of the informants themselves. It is in this sense that the intention behind the preliminary stages of scale development is carried forward into the analysis of the main survey data.

In the above example the process of seeking the correlates of the two scales in order to validate them started with hypotheses based on the interpretation of scale content but led to a network of correlates which in fact came to constitute a theory about the phenomena. There was a process of successive hypothesizing and checking which eventually came to rest on inferences from the entire network. Most of the other specific attitudes measured in the survey, as listed above, were brought into the argument. Some of the more important correlates in the argument were with the smoking-need scales, whose validity depends in turn on correlations with other parts of the data. These smoking-need scales were constructed by a process similar to that for the attitudes. Rather than the validity of an attitude scale, it thus becomes meaningful to speak of its usefulness as part of the total model or network built up in the process of validation. A scale's validity rests finally with the viability of this model as a source of testable propositions.

7.6 DISCUSSION

7.6.1 Relationship of the Alpha Approach to Other Techniques

Although complex psychometric issues arise under this head, there are some outstanding points that can be made briefly.

(a) There is a direct relationship between item selection procedures on the alpha approach and the Likert technique. The latter selects items which have the highest value for the correlation, r_{it}, between the score on the item and the total score on the set of items. This approach maximizes internal consistency, as defined by $\bar{r}_{it}$ the average of all such correlations.

It can be shown, (e.g. Nunnally, 1967) that

$$\bar{r}_{ij} = \bar{r}_{it}^{\,2} \tag{2}$$

when all item intercorrelations are equal, a condition which is approximated for a single cluster of items. The selection of items which maximizes $\bar{r}_{ij}$ on the alpha approach, and $\bar{r}_{it}$ on the Likert techniques, would therefore tend to be the same. Substituting (2) in (1) an alternative formula for alpha is obtained which could be used with the Likert technique. It is not so convenient as the $\bar{r}_{ij}$ formula for monitoring the effects of item selection on reliability, since each time an item is discarded it is necessary to return to the raw data to find the r_{it} values for computing the revised alpha estimates.

The principal difference between the alpha and Likert procedures is the inability of the latter to detect when more than one dimension is represented in the item set. Consider the r_{it} values in the last column of Table 7.4, for example. No correction has been made for the inclusion of the item itself in the total score; this does not affect the point being made. The values give no indication that the item set divides into two distinct clusters. Described by Green (1954) as a single common factor model for attitude measurement, the Likert technique at best will lead to a scale which measures what is most general to an item set. As a tool for exploring and representing the structure of attitudes in a population it therefore has limitations. Distinct attitudes of importance to substantial numbers of informants may well be missed in the process of imposing a hypothesis of undimensionality on the data. Usually the separate clusters in a matrix will overlap, as in Table 7.3, so that the hypothesis would be not so much false as oversimple. The technique however contains no safeguard against the inadvertent scaling together of clusters of items which are independent. The composite score for such an item combination would be a confounding of the different dimensions present and hence uninterpretable.

(b) The value of $\bar{r}_{ij}$ is closely related to the coefficient of reproducibility in Guttman or cumulative scaling (White and Saltz, 1957; Cronbach, 1951; Scott, 1960). A perfect Guttman scale would also be a perfectly reliable scale in terms of alpha. Perfect or even near-perfect scales hardly ever occur in practice. Both item unreliability and the presence of more than one dimension can lead to low reproducibility, but the Guttman technique does not allow these quite different sources of 'error' to be distinguished. It is a technique therefore for testing a hypothesis of unidimensionality rather than for exploring the number of dimensions present. Items covering two independent dimensions would not Guttman scale, but two distinct but related dimensions might. A cumulative scale with a coefficient of reproducibility of 0.91 can in fact be constructed from items 3, 4, 7, 9, 10, 11, 12, (which is the cumulative order) in Table 7.3. Like the Likert procedure Guttman scaling focuses on what is most general in the item set at the risk of missing important secondary dimensions. But Guttman scaling, even when applied to the items in a single cluster, does not afford a criterion for assessing how far a lengthening of the scale by adding items will improve its efficiency as a measuring instrument. Nor does it provide clearcut criteria for the selection of items. Certainly, the Guttman model makes fewer assumptions about the data, in that only relations at an ordinal level of measurement are assumed. But any advantage here may be thought dearly bought at the expense of the practical limitations just noted. The issue of measurement levels is discussed more generally below.

(c) The present alpha approach, being based on the values of r_{ij}, is easily combined with factor and cluster analysis techniques. So combined there is a ready safeguard against false inferences of unidimensionality, and an additional gain in the information obtained about reliability in relation to test length.

Guttman scaling, the Likert technique and factor and cluster analysis are all internal consistency approaches to the problem of scale construction. The researcher who uses any of these techniques will be protected against using ostensible measures that are actually of zero or near zero reliability. Factor and cluster analysis in addition provide information on the number of separate dimensions in the item pool. But the alpha approach, especially when combined with factor or cluster analysis, does all these things, and in addition proves a simultaneous assessment of reliability in relation to test length as well as providing a criterion for item selection.

(d) In the Likert technique and its extension into the alpha approach, evidence for the dimensionality of the attitude domain emerges directly from the covariance found among the responses informants give when reacting straightforwardly to opinion statements in terms of their own attitudes. The same is true, essentially, for Guttman scaling. The Thurstone or Judgement method is based upon an entirely different algorithm in this respect, since the original judges are asked to respond in terms of the assumed underlying continuum, independently of their own attitudes. The technique therefore carries a built-in bias towards the initial assumption that the actual attitudes of respondents can be adequately represented by a single dimension. The degree to which this is so is not directly tested. More than in the other methods, *a priori* notions of unidimensionality are imposed on the attitude domain, since items are scaled only by the direct assessments of a group of judges operating on this belief. Nor does the method lend itself readily to any mechanism for exploring salient components or subfactors in respondents' attitudes. The Thurstone technique was never of course intended as a tool for exploratory attitude surveys, and for the above reasons it remains difficult to adapt for this purpose.

Thurstone (1928) originally introduced a 'criterion of irrelevance', seldom used, but intended to be applied at a second stage to data collected from respondents. The criterion is not dissimilar in principle from the logic of item analysis used in Likert scaling. Others (Edwards and Kilpatrick, 1948, and Eysenck and Crown, 1949) have made more systematic attempts to synthesize the methods of item selection and scoring in the Likert and Thurstone approaches. The possibility of combining factor analysis with Thurstone scaling has been mooted even quite recently (Ostrom, 1971). But with the growth of understanding of the algorithms underlying these scaling models it became apparent fairly early that the Thurstone judgement type of models could not be combined with models based on response covariance, except by default. The more items approximate the model for

Thurstone scaling, in fact, the less suitable they are for correlational or factor analysis. The technical reason for this is, briefly, that Thurstone items have nonmonotone operating characteristics (see the discussion in Green (1954), pp. 343, 344, and 365).

If it is desired to dichotomize the population into the proportions *pro* and *con* on an issue then the Thurstone technique provides a means of doing so since the split can be made at the neutral zone defined by the mid-point of the scale. The method is not entirely satisfactory for this purpose since there is the general difficulty, which becomes most acute in the case of neutral items, of formulating statements that have the exact properties required by the Thurstone model. The only other rational method for obtaining a zero point on an attitude scale is that developed by Suchman (1950), and involves asking an intensity question after each attitude question. The development was in terms of cumulative scaling but is applicable to any attitude scale. Suchman also showed that often, but unfortunately not invariably, it was possible using a special 'foldover' technique to abstract the intensity information by the same question, thereby avoiding the doubling of the part of the questionnaire devoted to a particular area. Although the problem would seem to be of great importance for obtaining, say, an unbiased measure of the weight of public opinion for or against an issue, little progress has been made on the operational definition of a neutral point on an attitude scale since Suchman's poineering work. This may be because by and large those researchers who have been most concerned to use rigorously defensible scaling techniques have been seeking only the sources or correlates of individual differences in attitudes. For this purpose information on the relative position of respondents is all that is required, and an absolute measure is unnecessary.

(e) Attitudes vary in the degree to which they are general or specific, ranging all the way from highly abstract or ideological domains such as alienation, authoritarianism or radicalism—conservatism at one extreme, down to attitudes towards very specific stimulus objects such as an occupation or political party or issue at the other. A virtue of Likert-type procedures is that they can be adapted to attitudes of any degree of generality or specificity. For a discussion of the various considerations that arise in the measurement of attitudes that are either highly specific or highly general see McKennell (1974b). In general, homogeneity as indicated by the average interitem correlation $\bar{r}_{ij}$ is reduced as the domain becomes more abstract. More items are then required to attain a given reliability level. Thurstone-type techniques run into special difficulties when the attitudinal concept the judges are required to operationalize has to be defined at a high level of abstraction. The Guttman method is even more limited to attitudes of the specific kind, and the more so as one adheres to Guttman's criterion of tight cumulative interdependence between the items (high reproducibility).

7.6.2 Item Scaling

All the techniques so far considered can be said to be respondent-centred, as are most cross-sectional surveys of attitudes, in that the primary objective is the scaling of the respondents rather than the items, which merely serve as stimuli for this purpose. 'Stimulus-centred' techniques, of which only the briefest account can be given here, become necessary where the interest is less in the comparison of respondents than of the stimuli themselves. In the more common form of the problem what is required is a perceptual and/or preference mapping of several specific stimulus objects in the same logical class, e.g. products or brands, television programmes, occupations, politicians, etc. Until recently this class of attitudinal problem has been handled almost exclusively by attribute sorting techniques of which the semantic differential is the best known variant. For further discussion and references, see McKennell (1974b, pp. 64–66). Briefly, the stimulus objects are rated against a set of rating scales. The scales are then factor analysed after which the concepts can be plotted, for the entire sample or sections of it, in the space defined by the factor axes.

A point about the semantic differential approach to stimulus mapping is that, like the Likert and alpha techniques and mental test theory in general, it rests on untested interval scale assumptions. Recent years have seen a growing interest in models of representational measurement for which the axiomatic properties of the underlying metric are most stringently articulated. Yet where the data are such that both metric and non-metric mapping approaches can be applied they appear as often as not to yield similar stimulus configurations. The conditions where they do not, or where one approach is superior to the other, remain obscure. More empirical comparisons, such as that by Green (1969), are required.

When the stimuli to be mapped are attitude statements, and an interitem correlation matrix provides the input data, then both nonmetric methods and cluster/factor analysis with alpha scaling can be applied to the same data. Comparative data are lacking, but a technique like INDSCAL, as discussed in Chapter 6, for example, would appear to afford a more elegant means for mapping the items and certainly for identifying and summarizing group differences in 'factor' structure. The psychometric test model, on the other hand, would appear to yield dimensions that are more interpretable and deployable in variable form and, above all, it provides a rationale, through the reliability concept, for selecting the minimum numbers of items to be used as variables. It is the latter aspect which enables the notorious unreliability of responses to single attitude statements to be taken into account without resorting to impractically long batteries of items at the main survey stage. One suspects that where data exhibit strong structure this will emerge on either metric or non-metric approaches, while in other respects the special advantages and limitations of either approach may prove complementary. For further discussion of the metric question see the end of the chapter, also McKennell (1974b). In the latter reference it is argued that the securing of adequate inputs into the initial item pool which provided the basis for scaling is more important for representing attitude structures than are efforts at further refinement of the basic metric. At the present time it is not possible to offer more adequate guidelines than this on the choice between metric and non-metric models. What does seem important is that the choice should not be made on the basis of *a priori* preferences. Practitioners should be able to choose on the basis of what works best in practice, and for this more comparative data are required. The extent to which an approach does or does not make assumptions about the 'level of measurement', it is suggested, need not and ought not to be the overriding criterion guiding its choice. We end the chapter on this theme.

7.6.3 Levels of Measurement

The alpha approach, stemming from psychological test theory, entails manipulations of the data which assume interval scale properties even at the item level. The calculation and interpretation of $\bar{r}_{ij}$ as a product moment correlation, which the theory requires, for example, involves this assumption (amongst others). The approach therefore runs counter to the convention that has grown up in both psychology and sociology (stronger perhaps in the latter) that statistics appropriate for interval measures are inadmissable with ordinal data. This convention has so entered the folklore of contemporary social science that many seem unaware of its controversial status. The standpoint adopted by this author is pragmatic. For a fuller expression of the pragmatic approach to measurement see McKennell (1974b, pp. 11–13). Measurement conventions are not immutable. This particular one, though venerable, has never been universally accepted, and there is currently some reaction against it. By taking a stronger view of the data, access is gained to the world of measurement of the psychometrician. More powerful analytical techniques become available. The burden of this chapter has been that these psychometric techniques, when suitably adapted, are extremely useful for scale construction in the sample survey context. At the present stage of social science, it can be argued, fruitfulness in applied research is as good a guide to the choice of measurement model as the views taken by rival theorists about its axiomatic properties.

REFERENCES

Alwin, D. F. (1974), 'Approaches to the interpretation of relationships in the multitrait-multimethod matrix' in *Sociological Methodology 1973–1974*, H.L. Costner (Ed.), San Francisco: Jossey Bass.

Bonjean, C. R., Hill, R., and McLemore, S. (1968), *Sociological Measurement*. San Francisco: Chandler Pub. Co.

Borgatta, E. F. and Stolz, W. (1963), 'A note on a computer program for a rearrangement of matrices', *Sociometry*, **26**, 391–392.

Borgatta, E. F. (1965), 'Difficulty factors and the use of phi', *J. Gen. Psychol.*, **73**, 321–337.

Brannon, R. *et al.*, (1973), 'Attitudes and action: a field experiment in a natural population', *Am. Sociol. Rev.*, **38** 625–636.

Campbell, D. T. and Fiske, D. W. (1969), 'Convergent and discriminant validation by the multitrait-multimethod matrix', *Psychological Bulletin*, **56**, 81–105.

Cronbach, L. J. (1951), 'Coefficient alpha and the internal structure of tests', *Psychometrika*, **16**, 297–334.

Cronbach, L. J. and Meehl, J. (1955), 'Construct validity in psychological tests', *Psychol. Bull.*, **52**, 177–193.

Dingman, H. F. (1958), 'The relation between coefficients of correlation and difficulty factors', *Brit. J. Stats. Psychol.*, **11**, 13–17.

Edwards, A. L. and Kilpatrick, F. P. (1968), 'A technique for the construction of attitude scales', *J. Appl. Psychol.*, **32**, 374–384.

Edwards, A. L. (1957), *Techniques of Attitude Scale Construction.* New York: Appleton-Century-Crofts.

Eysenck, H. J. (1954), *The Psychology of Politics*, London: Routledge.

Eysenck, H. J. and Crown, S. (1949), 'An experimental study in opinion attitude methodology', *Int. J. Opin. Attit. Res.*, **3**, 47–86.

Festinger, L. (1947), 'The treatment of qualitative data by scale analysis', *Psychol. Bull.*, **44**, 146–161.

Frost, W. A. K. and Braine, R. L. (1967), 'The application of repertory grid technique problems in market research', *Commentary*, **9**, 161–175.

Goldthorpe, J. H., Lockwood, D., Bechhofer, F., and Platt, J. (1968), *The Affluent Worker: Industrial Attitudes and Behaviour*. Cambridge: University Press.

Green, B. F. (1954), 'Attitude measurement', in G. Lindzey (Ed.), *Handbook of Social Psychology*. Reading, Mass: Addison-Wesley.

Green, P. E. (1969), 'Dimensional interpretation and configuration invariance in multidimensional scaling: an empirical study', *Multivariate Behavioral Research*, **4**, 159–180.

Guttman, L. (1944), 'A basis for scaling qualitative data', *Am. Sociol. Rev.*, **9**, 139–150.

Guttman, L. (1950), in S. A. Stouffer *et. al.*, *Measurement and Prediction*, Vol. 4. New York: John Wiley.

Harman, H. H. (1967), *Modern Factor Analysis*, (2nd. ed.), Chicago: University of Chicago Press.

Hauser, M. and Goldberger, A. S. (1971), 'The treatment of unobserved variables in path analysis', in H. L. Costner (Ed.) *Sociological Methodology 1971*, San Francisco: Jossey-Bass.

Hendrickson, A. E. and White, P. O. (1966), 'Higher-order factor rotation', *British Journal of Mathematical Psychology*, **19**, 97–103.

Horn, J. L. (1969), 'On the internal consistency reliability of factors', *Multivariate Behavioral Research*, **4**, 115–125.

Horn, J. L. (1965), 'An empirical comparison of various methods for estimating common factor scores', *Educ. and Psychol. Measmt.*, **25**, 313–322.

Hovland, C. I. and Sherif, M. (1952), 'Judgemental phenomena and scales of attitude measurement: item displacement in Thurstone scales', *Journal of Abnormal and Social Psychology*, **47**, 822–32

Hyman, H. H. (1955), *Survey Design and Analysis: Principles, Cases and Procedures,* Glencoe, Ill: Free Press.

Katz, D. (1944), 'The measurement of intensity', in H. Cantril (Ed.) *Gauging Public Opinion*, Princeton: Princeton Univ. Press.

Lemon, N. (1973), *Attitudes and their Measurement*, London: Batsford.

Komorita, S. S. and Graham, W. K. (1965), 'Number of scale points and the reliability of scales', *Educ. and Psychol. Measmt.*, **25**, 987–995.

Lazarsfeld, P. F. (1955), 'The interpretation of statistical relations as a research operation', in P. F. Lazarsfeld and M. Rosenberg (Eds.)*The Language of Social Research*. New York: Free Press.

Lazarsfeld, P. F. (1944), 'Controversy over detailed interviews – an offer for negotiation', *Public Opinion Quarterly*, **8**, 38–60.

Likert, R. (1932), 'A technique for the measurement of attitudes', *Archives of Psychology*, New York, No. 140.

Loevinger, J. (1948), 'The technique of homogeneous tests', *Psychological Bulletin*, **45**, 507–529.

Lord, F. M. and Novick, M. R. (1968), *Statistical Theories of Mental Test Scores*, Reading, Mass :Addison-Wesley.

Matell, M. S. and Jcoby, J. (1971), 'Is there an optimal number of alternatives for Likert scale items?', *Educational and Psychological Measurement*, **31**, 657–674.

McKennell, A. C. (1969), 'Methodological problems in a survey of aircraft noise annoyance', *The Statistician*, **19**, 1–29.

McKennell, A. C. (1967), 'Adults' and adolescents' smoking habits and attitudes', *H.M.S.O. Government Social Survey Report*, 353B.

McKennell, A. C. (1970), 'Attitude measurement: use of coefficient alpha with cluster or factor analysis', *Sociology*, **42**, 227–245.

McKennell, A. C. and Thomas, R. K. (1967), 'Smoking habits and attitudes of adolescents and adults', *H.M.S.O. Goverment Social Survey Report*, SS 353/B.

McKennell A. C. (1974a), 'Surveying subjective welfare: strategies and methodological considerations', in *Subjective Elements of Well-being*, OECD:Paris (Burkhard Strumpel (Ed.)).

McKennell, A. C. (1974b), *Surveying Attitude Structures*, pp.45–55, Amsterdam: Elsevier.

McKennell, A. C. (1975), 'An application of Joreskog's general model to the analysis of subjective social indicators' *Unpublished paper* presented at British Sociological Association, Mathematics, Computing and Statistics Applications Group, Annual Conference, Canterbury.

McQuitty, L. L. (1957), 'Elementary linkage analysis', *Educ. and Psychol. Measmt.*, **17**, 207–229.

McQuitty, L. L. (1960), 'Hierarchical linkage analysis for the isolation of types', *Educ. and Psychol. Measmt.*, **20**, 55–67.

McNemar, Q. (1946), 'Opinion-attitude methodology', *Psychological Bulletin*, XLIII, 289–374.

Miller, Delbert C. (1970), *Handbook of Research Design and Social Measurement*, 2nd. Ed. New York: David McKay Co. Inc.

Moser, C. A. and Kalton, G. (1971), *Survey Methods in Social Investigation*. London: Heinmann.

Nunnally, J. C. (1967), *Psychometric Theory*, New York: McGraw-Hill.

Ostrom, T. M. (1971), 'Item construction in attitude measurement', *Public Opinion Quarterly*, **35**, 593–600.

Richardson, M. W. (1941), 'The combination of measures', in P. Horst (Ed.) *The Prediction of Personal Adjustment*, New York: Social Science Research Council.

Robinson, J. P. *et.al.* (1968–69), *Measures of Political Attitudes: Measures of Occupational Attitudes*, and *Measures of Social Psychological Attitudes*. Ann Arbor: Institute for Social Research.

Rosenberg, M. (1968), *The Logic of Survey Analysis*, London: Basic Books.

Schmitt, J. and Leiman, J. M. (1957), 'The development of hierarchical factor solutions', *Psychometrika*, **22**, 53–61.

Schooler, C. (1969), 'A note of extreme caution on the use of Guttman Scales', *Am. J. of Sociol.*, **74**, 296–301.

Schuman, H. and Johnson, M. P. (1976), 'Attitudes and behaviour', *Annual Review of Sociology*, in the press, to be published.

Schuman, S. and Duncan, C. D. (1974), 'Questions about attitude survey questions', in *Sociological Methodology 1973–4*, H. L. Costner (Ed.).

Scott, W. A. (1960), 'Measures of test homogeneity', *Educ. and Psychol. Measmt.*, **20**, 751–757.

Scott, W. A. (1968), 'Attitude measurement', in Vol. 2. G. Lindzey and E. Aronson (Eds.) *The Handbook of Social Psychology*, Second Edition, London: Addison-Wesley.

Seiler, Lauren H. and Hough, Richard L. (1970), 'Empirical comparisons of the Thurstone and Likert techniques', Chap. 8 in *Attitude Measurement*, G. F. Summers (Ed.), Chicago: Rand McNally and Co.

Shaw, Marvin E. and Wright, Jack M. (1967), *Scales for the Measurement of Attitudes*, New York: McGraw-Hill.

Suchman, E. A. (1950), 'The intensity component', in S. A. Stouffer *et. al. Measurement and Prediction*, Princeton: Princeton University Press.

Thurstone, L. L. (1928), 'Attitudes can be measured', *Am. J. Sociol*, **33**, 529–554.

Tryon, R. C. (1957), 'Reliability and behaviour domain validity: reformulation and historical critique', *Psychol. Bull*, **LIV**, 9–17.

Upshaw, H. S. (1968), 'Attitude measurement', in H. M. Blalock and A. M. Blalock (Eds.), *Methodology in Social Research*, New York: McGraw-Hill.

White, W. B. and Saltz, E. (1957), 'Measurement of reproducibility', *Psychol. Bull*, **54**, 81–99.

Chapter 8

Binary Segmentation: The Automatic Interaction Detector and Related Techniques for Exploring Data Structure

*A. Fielding**

8.1 INTRODUCTION

Statistical methodology for dealing with relationships between variables has largely been developed, until recently, in response to the data analysis needs of the experimental scientist. Controlled experimentation has led, except at the very elementary level, to a study of the structure and pattern of relationships between variables by statistical inference in the context of explicit or implicit statistical models. Interest in such structures has been concentrated on parameter estimation, theory and model testing, in parallel with heated controversies about the appropriate forms of inference to use. Yet the empirical tool of the vast majority of social researchers is the sample survey, characterized by lack of control over key variables and resulting in a complex, and often large, set of data. Here, the approach to data analysis must often be exploratory rather than inferential. The requirements are for techniques which adequately summarize and describe structural dependencies in the data set and which are aimed at uncovering empirical regularities. These form a prelude to theory generation and model construction.

For these purposes, of course, the traditional body of descriptive statistical methods is not entirely inappropriate. Tables of means, two and three way crosstabulations, low order correlations, simple graphs and diagrams and the like, are clearly indicative of certain aspects of data structure. However, there is a limit to the researcher's capabilities in this direction, with large data sets and numerous relevant variables. In many situations methods are required to formalize the search for patterns in data which are then used to formulate the structure of relationships between variables. Before the advent of computers this 'ransacking' or 'data dredging' task of examining the relationships of the main variables under study with a host of others taken singly or in combination was carried out on an *ad hoc*

*I would like to thank R. Silverman of the University of Bristol for making available his data collected as part of a project in which the author was a statistical adviser.

basis and proved time consuming. Lazarsfeld's method of elaboration and specification, for instance, was an early example of this approach in a search for structural relationships in categorical data (Lazarsfeld, 1955). When the logic of this method is examined it is apparent that it is practicable only for situations where a small number of variables is involved. The computer based binary segmentation procedures are designed to overcome these difficulties.

The operation of this class of procedures on a data set may be made clear by stating formally the principles underlying it.

A group of n individuals or objects is under study each characterized by a set of variables $(y_1, y_2, \ldots, y_m; x_1, x_2, \ldots, x_p)$. The y_i are taken to be the dependent variables, x_j independent or predictor variables. Each of the variables may be measurements on any scale but more usually interval, ratio, ordered or a labelled categorization. In many contexts there is just one dependent variable y. We denote the set of values which each variable can take by Y_i or X_j. A binary segmentation induced by a variable x_j would be some way of splitting up the n individuals into two groups according to which of two disjoint but exhaustive subsets of X_j their value of x_j belonged. Thus, for an election voting behaviour variable with values in the set X_j/(Labour, Conservative, Liberal, Others), the two subsets (Labour), (Not Labour) could induce a binary split of individuals. For interval scale variables, such as height, X_j is a portion of the real line and the split $(X_j : X_j > 66$ ins) and $(X_j : X_j \leqslant 66$ ins) could induce a division of the individuals into two groups on the basis of their heights.

All the binary segmentation procedures to be discussed adopt this approach. Firstly the sample as a whole is split into two subgroups and the same procedure is applied sequentially to resulting subgroups. However, at any stage decisions must be made on the particular predictor variable to be used in inducing the split, and how the two subsets of its values are to be formed. Since the objective is to explain the variation in the dependent variables, it seems sensible to make these two decisions so that resulting subgroups are as different as possible and are internally as homogeneous as possible *vis-à-vis* the y_i's. The specific algorithms available within the general framework differ either in the level of data for which they are appropriate or in the way in which they define a criterion measuring subgroup separation on the dependent variable(s), or 'distance' functions as they are often called. These variants of the general procedure will be discussed in turn in Section 8.2. For all of them, however, the predictor variables must be either ordinal or nominal.

The basic strategy is to examine each predictor variable x_j and to choose a partition of its set of values such that the induced binary split results in subgroups that maximize the chosen 'distance' function. The next step is to choose among these the variable which produces the overall maximum. In a loose sense it is the 'best explanation' of the y_i we can get through a single split into two groups. The process continues by sequentially splitting the subgroups obtained from the previous stage or stages using the same criterion for deciding which predictor to use.

The result of the binary segmentation process is a hierarchical tree or subgroups a simple example of which is given in Figure 8.1. The dependent variable in this

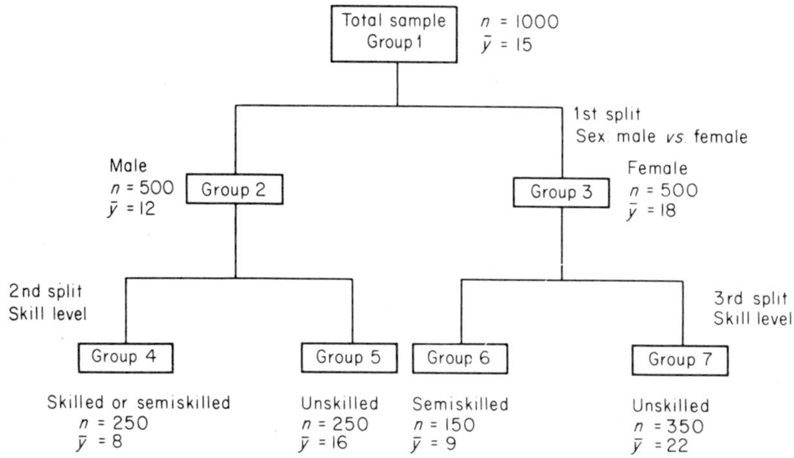

Figure 8.1. An example of a tree diagram. (Dependent variable: Absence from work (days per year). Predictors: Age, sex, skill level).

case is absence from work (in days per worker year) and is therefore an interval scale variable. The criterion used for splitting the samples is the between sum of squares on the dependent variable, which is discussed in detail in Section 2.

Only two of the three predictors are used in the diagram. The first split is on sex. Male workers have an absence rate of 12 days per year and female workers 18 days per year. Group 2 (male) is next split on level of skill, unskilled workers falling in one subgroup, skilled and semiskilled workers in the other. Group 3 (female) is also split on skill level and divided into semiskilled and unskilled workers.

The diagram gives us three types of information:

(i) *The terminal (final) groups.* The workers in the sample are classified into four groups – in ascending order of absence rates, they are (a) skilled and semiskilled male workers, (b) semiskilled female workers, (c) unskilled male workers and (d) unskilled female workers. These could provide the input for a predictive model.
(ii) *Association.* The two predictors sex and skill level are associated. A considerably higher proportion of the female workers are unskilled and this partly explains the difference in absence rates between men and women.
(iii) *Interaction*. The presence of interaction between sex and skill level is shown by the different effect of skill level for men and women. In the case of the men, the difference between absence rates is 8 days per year; for women, the difference is 13 days per year. Similarly, the sex effect for unskilled workers is 6 days per year; for the more skilled it is only 1 day per year.

A more detailed discussion of how such trees and intermediate output in the tree formation process can be interpreted is presented in Section 8.3.

The ultimate aim of the analysis may be to develop a predictive model for the dependent variable, but an initial exploration may answer a series of much 'looser' questions connected with the specification of structural dependencies. How is each predictor variable taken by itself related to the dependent variables? Does the specification of these effects explain a substantial proportion of the variation in the dependent variable? What is the effect of a particular predictor and how strong is it, after adjusting for the effects of and associations with other predictors? Here it is recognized that with predictors closely related to each other, the marginal effects of one may be partly subsumed in the marginal effects of others, and there may be a difficulty in disentangling them. Again it may be recognized that the effect of a particular variable may have different patterns for different levels of other variables.

Implicit in these loose questions and observations is a recognition of two distinct aspects of structural relationships: the problems caused by associations between predictors in specifying them, and how predictors interact with each other in affecting the dependent variables.

As has been emphasized the traditional approaches to such data analysis has been in the context of models. Possible interaction effects are incorporated in a model, the parameters estimated and tested for significance. The model is the framework; the inference is easier if balanced orthogonal designs of experimental data are available. All this requires specification in parametric form of the nature of the effects. Even the recent developments of log-linear models (see Chapter 4 of Volume 2) for categorical predictors require a detailed parametric model specified at the outset, which must be assumed to be appropriate. In any event a full specification of such a model for many predictors would require a large amount of data. If data are sparse undue restrictions are placed on the parameters so that even if a full and well specified model is appropriate for the structure at hand, it becomes unrealistic. What is necessary in exploratory research is an approach which is free from such restrictions, but which sets about an examination of the data in a formal though flexible way. The emphasis shifts away from models and restrictive parametric forms as a starting point to exploration with the aim of ensuring that such models when suggested and developed will be more relevant.

Many such approaches have been labelled 'data dredging' and derided. This is unfortunate since the exploratory phase and inferential phase act in concord. Data exploration suggests hypotheses. In turn formal hypotheses when tested may suggest more illuminating ways of examining the data. One possible reason for the low esteem of exploratory analysis is the widespread use of techniques such as factor analysis as data-dredge, in situations where the implicit models underlying the technique is inappropriate. The availability of large scale computing facilities and a proliferation of little understood packages has encouraged this. Yet the disadvantages of such facilities in this context become advantages when the aim is the development of exploratory approaches which are free from implicit model restraints. They have provided a medium by which it is possible to operate formal procedures in algorithmic form which attempt to simulate the activity of a social researcher in examining the structural relationships between variables in a large data set. The end result is a systematic rather than haphazard approach to data summary which facilitates directed rather than undirected explorations.

Binary segmentation is one such approach which does this in a simple, intuitive, though fruitful way. It attempts to produce a hierarchical data structure and meaningful intermediate results which guide the researcher in his formulation of appropriate hypotheses about patterns of effect.

8.2 BINARY SEGMENTATION AS A GENERAL METHOD

Several automatic procedures are now available in a form designed for computer implementation. The principal cases are described below. The basic objective of the procedures is to ascertain how the variation in the dependent variables is related to or explained by the predictors. Therefore it is appropriate that the splitting process should operate so that within the subgroups resulting from a particular split the elements should be as similar as possible in terms of the scores on the dependent variables and that the difference between the subgroups should be maximized.

Case 1: One dependent variable, interval scale

The Automatic Interaction Detector (AID) is a label given to a series of algorithms developed by the Institute for Social Research, University of Michigan (Morgan and Sonquist, 1963; Sonquist and Morgan, 1964; Sonquist, 1970; Sonquist, Baker and Morgan, 1971). An automatic procedure which is essentially the same has also been programmed independently by Gillo (1972). It deals with the simplest case of one interval scale dependent variable y. Here there is little disagreement over the choice of an appropriate distance function or criterion to optimise in achieving maximal subgroup separation on y. Since y can be thought of as taking values which are real numbers, a distance between two values Y_l, Y_k may be based simply on $Y_l - Y_k$. To make the distance measure more suitable for generalizing to higher dimensions it is usual to operate on $(Y_l - Y_k)^2$, the usual form of Euclidean distance. In the case of the single dimension under consideration, of course, maximizing such squared distances is equivalent to maximizing absolute distances. With these distance measures in mind it is easy to draw a parallel between the AID criterion and the familiar between group sum of squares in the analysis of variance.

The following notation is used:

$\bar{Y}$ is the mean of the parent group of size n.

$\bar{Y}_1$, $\bar{Y}_2$ are the subgroup means resulting from a binary split induced by a segmentation of the values of some predictor variable x_j, into groups of size n_1, n_2 respectively.

Y_{1k} $(k = 1, 2, \ldots, n_1)$, $Y_{2k} (k = 1, 2, \ldots, n_2)$ are the observations on the dependent variable in each group.

Three quantities can be defined:

$$n_1(\bar{Y}_1 - \bar{Y})^2 + n_2(\bar{Y}_2 - \bar{Y})^2 = \frac{n_1 n_2}{n}(\bar{Y}_1 - \bar{Y}_2)^2 \qquad (1)$$

$$\sum_{k=1}^{n_1} (Y_{1k} - \bar{Y}_1)^2 + \sum_{k=2}^{n_2} (Y_{2k} - \bar{Y})^2 = \sum_{l=1}^{2} \sum_{k=1}^{nl} (Y_{lk} - \bar{Y}_l)^2 \qquad (2)$$

$$\sum_{l=1}^{2} \sum_{k=1}^{nl} (Y_{lk} - \bar{Y})^2 \qquad (3)$$

The expression (1) is the frequency weighted sum of squares of deviations of group means from the parent mean, (2) is the sum of subgroup heterogeneities, and (3) is heterogeneity in the parent group.

Heterogeneity is defined by the usual variation criterion in analysis of variance: the sum of squared distances of individual y values from the mean. The smaller this quantity the more homogeneous a group. It will be seen that the distance between subgroups which is maximized is the usual between group sum of squares for two groups, the frequency weighted sum of squares of deviations of subgroup means from the parent mean.

In the operation of the AID algorithm, the binary split actually induced is the one that maximizes the distance between group criterion (1), and minimizes (2). The former is of course equivalent to the latter since the sum of (1) and (2) is fixed and equal to the parent heterogeneity (3), the total sum of squares.

This criterion of distance is used to maximize the explanatory power of a binary split at each stage, i.e., the between group sum of squares (BSS) in the ordinary analysis of variance sense. On dividing the BSS (1) by the total sum of squares (TSS) or (3) we get Fisher's univariate correlation ratio for a split,

$$\eta^2 = \frac{\text{BSS}}{\text{TSS}} = \frac{n_1(Y_1 - \bar{Y})^2 + n_2(\bar{Y}_2 - \bar{Y})^2}{\sum_{l=1}^{2} \sum_{k=1}^{n_l} (Y_{lk} - \bar{Y})^2}$$

so, equivalently, maximizing the distance function maximizes the proportion of variation in y explained by the binary split. Some writers have suggested dividing the quantities by n, the overall group size, to get a more meaningful measure of average distance or variance, but since this is constant for a given group the optimizing process will not be affected.

There is an additional feature which makes this distance criterion a useful one. It may be thought that an examination of the binary splits induced by all $2^{k-1}-1$ possible partitions of the k categories of a predictor x_j is necessary to find the optimum. This would prove cumbersome even for relatively small k. However, it has been shown by Fisher (1958), that only $k-1$ partitions need be examined. If, for a parent group, k elements of X_j characterize individuals, one arranges the k groups in order of their means on the dependent variable. The optimal binary split for that particular predictor is then induced by one of the $k-1$ contiguous partitions defined on that order. The AID series of algorithms have a further feature which is used when predictors are only allowed to induce 'monotone' splits. If there is an ordering of the categories of a predictor and it is felt that y should be monotonically related to it, i.e. as x_j increases y should *a priori* increase, then the permissible splits are restricted. For the set X_j of ordered categories $\{1,2,3,4,5,\}$ a split induced by $\{1,2,3\}$ and $\{4,5,\}$ would be permissible whereas $\{1,2,4\}$ and $\{3,5\}$ would not.

Section 8.3 will discuss at some length the ways in which intermediate output and distance measures in addition to the tree structure itself for the AID algorithm can aid exploratory research.

Case II: More than one dependent variable, interval scale

A natural extension of the AID distance criterion is based on squared interpoint distance in m-dimensional space and is used in many divisive cluster algorithms (e.g. Edwards and Cavalli-Sforza, 1965). Here individuals may be characterized by m-element column vectors of values on the m dependent or criterion variables. A number m of each of the expressions (1), (2) and (3) can be formed, one for each variable, and expressed as vectors of parent group heterogeneity, pooled subgroup heterogeneity, and distances between subgroups. The sum of the elements of the vectors represents the extension to m-dimensions of the distance and heterogeneity criteria of the AID algorithms. Such a procedure has been developed in algorithm form as one of the optional distance measures in MAID-M (Gillo and Shelly, 1974).

The distance and heterogeneity criteria may be formally expressed in matrix terminology. Let $\bar{Y}_i$ be the parent mean of variable y_i, with respective subgroup means $\bar{Y}_{1i}$, $\bar{Y}_{2i}$. An individual's score on y_i is characterized by Y_{kli} with k determined by the group to which the binary split allocates the individual ($i = 1, 2, \ldots, m; k = 1,2; l = 1,2, \ldots, n_k$).

Let

$$t_{ij} = \sum_k \sum_l (Y_{kli} - \bar{Y}_i)(Y_{klj} - \bar{Y}_j),$$

$$w_{ij} = \sum_k \sum_l (Y_{kli} - \bar{Y}_{ki})(Y_{klj} - \bar{Y}_{kj}),$$

$$b_{ij} = \sum_k n_k(\bar{Y}_{ki} - \bar{Y}_i)(\bar{Y}_{kj} - \bar{Y}_j), \quad i,j = 1,2, \ldots, m$$

or equivalently

$$b_{ij} = \frac{n_1 n_2}{n}(\bar{Y}_{1i} - \bar{Y}_{2i})(\bar{Y}_{1j} - \bar{Y}_{2j})$$

A matrix $\mathbf{T}$ with elements t_{ij} is the total sum of squares and crossproducts matrix for the parent group. The matrix $\mathbf{W}$ with elements w_{ij} is the pooled within subgroup sum of squares and crossproducts matrix formed by the binary split. The matrix $\mathbf{B}$ with elements b_{ij} is the between subgroup sum of squares and crossproducts matrix. It can be shown, with the usual analysis of variance partitioning, that

$$\mathbf{B} + \mathbf{W} = \mathbf{T}$$

The distance function in the simple extension of AID to m dependent variables is Trace($\mathbf{B}$) and the binary split maximizing this at each stage is sought. The extension of the correlation ratio indicating the explanatory power of the split is

$$\eta_m^2 = \text{Trace}(\mathbf{B})/\text{Trace}(\mathbf{T})$$

which can be interpreted as the proportion of variation explained.

It may be noted that the order of values of this distance criterion is not invariant with respect to changes of scale of the y variables. The optimization process may then produce different binary splits for changes in the units of measurement of these variables. It is clear that if variables are in widely different units (e.g. age in years, weight in lbs), it might be preferable to standardize them before computing this distance criterion. Conversely if the variables are reasonably commensurate and measured on the same scale the differential variances of the variables may well be important and unstandardized variables may be used. Either way the researcher must satisfy himself that his scales are appropriate.

A closely related problem is that the variables will usually be correlated between individuals in a group. Such correlations can be eliminated by an orthogonal transformation of the variables, say to principal components. This rotation of the reference frame of the axes leaves the relative position of points in *m*-space unchanged, and hence distances will be invariant. However, it may be desired to standardize the resultant variables for variance. This is equivalent to changing the scales along the new axes so that all new variables have unit variance. This technique of transforming to a set of uncorrelated standardized variables is well known (e.g. Rao, 1965). Such transformations consist of an orthogonal rotation of the axes, followed by a change of scales along the new axes. The old axes thus become oblique to one another and all distances between individuals and hence the optimization process are in general affected. This transformation can be equivalently effected by postmultiplying the row vector of dependent variable values by a matrix $\mathbf{A}$, where $\mathbf{A}$ is such that $\mathbf{AA}' = \mathbf{V}^{-1}$ the inverse of the variance–covariance matrix of the *y*-values. It has usually been suggested that for use in binary segmentation procedures, $\mathbf{V}^{-1}$ be estimated from the pooled sums of squares and cross-products within the two groups formed by the binary split under consideration. It is proposed in this form by Bouroche and Tennenhaus (1972) who use as a distance criterion what is essentially Mahalonobis' D^2 (see Kendall and Stuart, 1968, p. 259). In a slightly different form it is embodied as an alternative distance criterion in MAID-M (Gillo and Shelley, 1964).

These suggestions may be developed more formally in matrix terms. Suppose $\mathbf{y}$ represents an *m*-column vector of dependent variable values with an estimated variance–covariance matrix $\mathbf{V}$. Let Λ be the diagonal matrix of eigenvalues of $\mathbf{V}$, aranged in order of size, and $\mathbf{C}$ the matrix having as its columns the corresponding eigenvectors. An orthogonal transformation of y to principal components followed by rescaling to unit variance variables results in a column vector $\mathbf{z}$ of new variables, such that

$$\mathbf{z}' = \mathbf{y}'\mathbf{C}\Lambda^{-1/2}$$

From the properties of eigenvalues and eigenvectors

$$\mathbf{V} = \mathbf{C}\Lambda\mathbf{C}'$$

so

$$\mathbf{V}^{-1} = (\mathbf{C}')^{-1}\Lambda^{-1}\mathbf{C}^{-1} = \mathbf{C}\Lambda^{-1}\mathbf{C}' \quad \text{since } \mathbf{C} \text{ is orthogonal.}$$

Choosing $\mathbf{A} = \mathbf{C}\Lambda^{-1/2}$ so that $\mathbf{AA}' = \mathbf{V}^{-1}$ the transformation becomes

$$\mathbf{z}' = \mathbf{y}'\mathbf{A}.$$

Denoting the total sum of squares and cross products of the new variables by $\mathbf{T}_z$, the between group sum of squares after a binary split by $\mathbf{B}_z$, and the pooled within subgroup sum of squares and cross products as $\mathbf{W}_z$ so that as before

$$\mathbf{T}_z = \mathbf{B}_z + \mathbf{W}_z$$

or in terms of the previously defined matrices for the y variables

$$\mathbf{TV}^{-1} = \mathbf{BV}^{-1} + \mathbf{WV}^{-1},$$

the criterion to be optimized becomes Trace $(\mathbf{BV}^{-1})$.

Actually the MAID-M algorithm adopts as the criterion Trace $(\mathbf{BW}^{-1})$, but if $\mathbf{V}'$ is defined from the pooled subgroup matrix $\mathbf{W}$, they differ only by a multiplying constant and the optimization process is unaffected.

Case III: One dependent variable, nominal scale

Binary segmentation methods appropriate for one categorized dependent variable and a series of categorized predictors are embodied in algorithms referred to as THAID (Messenger and Mandell, 1972; Morgan and Messenger, 1973). Here if the dependent variable has m categories it is useful to characterize an individual's y value if he belongs to category j by an m-element vector with zeros everywhere except for the jth element which takes value 1. The centroid of a parent group is $(p_1, p_2, \ldots, p_m)'$ where p_j is the relative frequency of individuals in category j. The centroids formed after a binary split are $(p_{11}, p_{12}, \ldots, p_{1m})'$ and $(p_{21}, p_{22}, \ldots, p_{2m})'$ with similar interpretations for the elements. It seems sensible to form a frequency weighted sum of 'distances' of the subgroup centroids from the parent group centroid as a criterion of subgroup separation. However, there is some arbitrariness about an 'appropriate' distance function to use. The THAID algorithm adopts an absolute distance criterion based on a city-block metric (see Johnson and Wall, 1969), although other algorithms use different distance metrics. The THAID criterion, delta, is defined as

$$\delta = n_1 \sum_{j=1}^{m} |p_{1j} - p_j| + n_2 \sum_{j=1}^{m} |p_{2j} - p_j|$$

or equivalently

$$\delta = 2\frac{n_1 n_2}{n} \sum_{j=1}^{m} |p_{1j} - p_{2j}|.$$

An induced binary split maximizing δ gives subgroups with maximal separation, in the sense that their relative frequency distributions have maximal absolute difference from that of the original subgroups, and hence from each other.

A similar criterion was earlier suggested by Belson for the simpler situation of a dichotomous dependent variable (Belson, 1959). Belson's aim was to generate variables, maximally predictive of the dichotomous response variable, which were to be used as matching criteria in an experiment or quasi-experiment. Though the motivation appears different the method again uses binary segmentation. The 'distance' criterion is referred to as 'maximum displacement'. The parent group of individuals can be cross-classified by a predictor and the dependent variable into a $k_i \times 2$ contingency table if the predictor x_i has k_i categories. Binary splits are induced by dividing the categories of x_i into two groups. Expected values of all frequencies under the independence hypothesis can be calculated in the usual way. Focusing on one of the two dependent variable categories, the difference between observed and expected frequencies for each of the k_i categories can be calculated. This is referred to as the displacement for that category. Some of these will be positive and others negative but their sum is zero. An optimum binary split is that induced by dividing x_i into those categories for which displacement is negative and those for which it is positive. The sum of the positive (or negative) displacements will be the total number of cases displaced and is a measure of the predictive power of that binary split. Clearly a split is made on the predictor for which this is a maximum. It can easily be shown that a binary split for a predictor maximizing displacement in Belson's terminology is equivalent to that maximizing the criterion δ for the special case $m = 2$.

With p_{1j}, p_{2j} and p_j defined as before:

$$\delta = \sum_{i=1}^{2} \sum_{j=1}^{m} | n_i p_{ij} - n_i p_j |$$

$$\delta = \sum_{i=1}^{2} \sum_{j=1}^{m} | O_{ij} - E_{ij} |$$

with $O_{ij} = n_i p_{ij}$, $E_{ij} = n_i p_j$ being the observed and expected absolute frequencies in the $2 \times m$ contingency table formed by the two subgroups and the m categories of the dependent variable. In the case of $m = 2$, the dependent variable dichotomy $|O_{ij} - E_{ij}|$ is the same for all four cells and can be labelled conveniently $|O - E|$. It follows that

$$\delta = 4 \, | O - E |$$

which is four times the displacement, in Belson's terminology. Maximizing displacement is thus equivalent to maximizing δ.

For this dichotomous dependent variable situation the 'distance' criterion used in the AID algorithm can also be used. Here y can be considered to take on values zero and unity. Means become proportions and in the terminology used above the criterion to be maximized becomes

$$n_1(p_{11} - p_1)^2 + n_2(p_{21} - p_1)^2$$

or

$$\frac{n_1 n_2}{n}(p_{11} - p_{21})^2$$

In comparing this with δ it will be seen that the criterion to be optimized involves squared rather than absolute distances and only one of two categories. Since relative frequencies are complementary in dichotomies this is equivalent to a 'distance' function using both. A useful comparison of the use of this criterion and δ applied to real data is given by Thompson (1972).

Squared distance has also been suggested as a criterion for group separation in the so called ELISEE method (Cellard, Labbé and Savitsky, 1967; Bouroche and Tennenhaus, 1972) for m-category dependent variables. Treating the vectors of relative frequencies as centroids, the distance between subgroups which is to be maximized is a frequency weighted sum of squared distances between subgroup centroids and parent centroids. Effectively this weights the squared differences of the relative frequencies by the inverse of the relative frequency in the parent group:

$$n_1 \sum_{j=1}^{m} \frac{(p_{1j} - p_j)^2}{p_j} + n_2 \sum_{j=1}^{m} \frac{(p_{2j} - p_j)^2}{p_j}$$

Actually the originators of the ELISEE method divide distance by n but this will obviously not affect the optimization process. It is readily seen that the above criterion is equal to the χ^2 statistic calculated from the $2 \times m$ table formed by the split, so it is at least an intuitively appealing criterion within a familiar context. Dividing by n gives χ^2/n which is Pearson's Q^2 measure of association in such tables. The criterion has also been suggested for use in divisive clustering when criterion variables are all dichotomies (McNaughton-Smith, 1963).

The simplest measure of subgroup distance for an m-category dependent variable is a natural analogue of the between group sum of squares used for m measured dependent variables:

$$n_1 \sum_{j=1}^{m} (p_{1j} - p_j)^2 + n_2 \sum_{j=1}^{m} (p_{2j} - p_j)^2$$

or equivalently

$$\frac{n_1 n_2}{n} \sum_{j=1}^{m} (p_{1j} - p_{2j})^2 \qquad (4)$$

This is in keeping with the tradition of multivariate one-way analysis of variance with a 'trace' metric or unweighted squared distance. The singularity of matrices of squares and cross products here prevents the usual approach to partitioning variation on the basis of determinants. The possibilities of using the criterion and the χ^2 *ELISEE* criterion are briefly discussed in the manual to THAID where neither is advocated (Messenger and Morgan, 1973). The reasons for this are not made clear.

An equivalent criterion has been suggested in the context of analysis of variance for categorical data (Light and Margolin, 1971). Actually, Light and Margolin's motivation was not binary segmentation and indeed their partition of sums of squares gives a criterion which is one-half the above.

These distance measures have focused on the differences in the relative frequency distributions of the dependent variable either in the absolute or squared distance form. However, the THAID series of algorithms does incorporate an additional criterion which may be useful. This is a prediction orientated measure with the same motivation as asymmetric prediction indices developed as measures of association in contingency tables (Goodman and Kruskal, 1954). In the absence of information on predictor variables the best prediction of the dependent variable category for an individual in a subgroup is the modal category for that subgroup. This gives the smallest probability of misclassification. If θ_y is the relative frequency in a parent group modal category then $(1-\theta_y)$ could be intuitively interpreted as its heterogeneity. A group's centroid vector can be defined to be the same as for individuals in the modal group (i.e. zero elements apart from the lth which equals unity, where l is the modal category). On taking an unweighted sum of the squared or absolute distances of the observation vectors from this it is seen that individuals in the modal category contribute zero and the $n(1-\theta_y)$ non-modal individuals contribute unity. Dividing by n gives $(1-\theta_y)$. Now suppose the dependent variable is given in each subgroup by the modal categories for the subgroups with relative frequencies θ_{y1}, θ_{y2} respectively. Since the relative frequencies in the two subgroups are n_1/n and n_2/n respectively the overall predictive power for this binary split is

$$\theta_y(x_j) = (n_1\theta_{y1} + n_2\theta_{y2})/n,$$

this being the overall probability of making a correct prediction given the particular binary split on x_j. The frequency weighted measure of heterogeneity within subgroups is $1-\theta_y(x_j)$. Between group 'distance' or separation could be defined as the difference between parent homogeneity and within group heterogeneity, $(\theta_y(x_j)-\theta_y)$, or the improvement in the probability of making a correct prediction induced by a binary split. Actually Goodman and Kruskal's λ would be $(\theta_y(x_j)-\theta_y)/(1-\theta_y)$, the relative improvement in predictive importance. The THAID algorithm operates on parent groups to maximize $\theta_y(x_j)$ but naturally the optimization process yields the same result whichever criterion is used. Morgan and Messenger (1973) have shown that the segmentation process using this criterion is likely to be somewhat different from that produced by other criteria. A decision must be made whether one really wants an optimum individual prediction model rather than one giving groups whose overall distributions on the dependent variable are maximally different. However, Gower (1971) has pointed out that in many circumstances this may well be what is implicit in the analyst's approach to his search strategy, though as with choices of measures of association there is an element of arbitrariness.

Case IV: One dependent variable, ordinal scale

In many circumstances the dependent variable may be an ordered categorization, and it seems desirable to use this additional information. One could score the categories and use the AID criterion, or ignore the ordinal information and treat y as if it were a nominal categorization. Clearly neither of these is entirely satisfactory, and a 'distance' function which uses the ordinal information has been suggested by Bouroche and Tennenhaus (1972). Again y is treated as a vector with m elements. However, if an individual belongs to the lth ordered category, elements of the vector up to the lth are one and the rest zero. The centroid vector for a group will be $(F(1), F(2), \ldots, F(m))$ where $F(l)$ is the cumulative relative frequency up to the lth ordered category. The between subgroup distance is

$$n_1 \sum_{j=1}^{m} \frac{F(j \mid 1) - F(j))^2}{F(j)} + n_2 \sum_{j=1}^{m} \frac{F(j \mid 2) - F(j))^2}{F(j)}$$

where $F(j|i)$, $i = 1, 2$ are the cumulative frequencies for the subgroups.

This is a frequency weighted sum of the difference between subgroup cumulative distributions and the parent distributions. If this criterion is maximized over all binary splits we clearly have two groups whose cumulative distributions are as different as possible. This directly uses the ordinality information contained in y.

Conclusions

There are, then, a whole series of criteria which could be used in an approach to the explanation of the dependent variable by a binary splitting process. Some have been incorporated in well-documented computer programs, in particular those in AID and THAID. The others could easily be developed into full scale algorithms. It has been noted that different levels of measurement of dependent variables dictate different approaches, although for each level there are a number of possibilities. This is in keeping with the usual controversy over what are appropriate 'distance metrics' for various types of data. Yet the usual end result of a binary segmentation process is a tree of subgroups with a great deal of summary information at each stage. It does at least provide in an intuitive and systematic way a framework for developing a 'picture' of the possible relationships between the predictor variables. Interpretation of the systematic summary may not always be easy, but using profiles of subgroup dependent variable means at each stage and distributions over predictor variable categories, and looking at when and how the binary splits occur, it is often possible to develop some fairly specific notions about the structure of the data set.

Binary segmentation proceeds by successively splitting groups at any stage into subgroups induced by the categories of some predictor. It is clear that the tree splitting process could theoretically continue until all configurations of the predictors are represented in the final subgroups of the tree. However, it is unlikely that the number of observations would permit representation of each configuration adequately. Even with a moderate number of predictors and categories in each, the number of total configurations would be very large, i.e.: $\prod_{i=1}^{p} k_i$ if p is the number of predictors with k_i categories for predictor i. Thus for four predictors with say 2,4,5,5 categories respectively this would be 200 configurations or potential final subgroups. But, even if the data were not 'sparse', to carry out the splitting process to its ultimate end would be wasteful and unnecessary. Usually the real effects on the dependent variable at late stages in the tree are insubstantial, if not spurious, and in any event the aim is to produce a systematic and economical summary of the data. Thus *stopping rules* are required to prevent further splitting of subgroups, if it is felt that such splitting would be inconsequential.

At present the stopping rules are based on rules of thumb which have appeared satisfactory in practice. However, it is apparent, as will be briefly discussed in Section 8.5, that from statistical and distributional considerations there may be scope for improvement in these. An obvious restriction is to prevent the development of very small subgroups, and all the algorithms currently available have a facility to prevent a group being split further unless the number of observations is greater than a level prespecified by the user. Also it is desirable that the explanatory power of the optimum binary split of a subgroup must be significant, if that group is to be split. In the AID algorithm this principle is operated in two ways. The ratio of the optimal BSS for a split to the TSS for the parent group must be greater than some level (.006 is the conventional level at present), otherwise it is felt that the explanatory power of a split is insubstantial. Also the total sum of squares for the parent group must be greater than a specified proportion of that for the overall survey group (0.01 is conventional). If this split eligibility criterion is not met the parent subgroup under consideration is regarded as homogeneous in the dependent variable. Similar criteria are required for the operation of the THAID algorithm. In the programmed versions of binary segmentation the process also stops if the number of subgroups reaches some prespecified level. The process must at any stage also decide which of the existing subgroups available for splitting will be next operated upon. In the AID implementation this is the group with the largest total sum of squares. In effect this is the subgroup with the largest potential for further explanation of the dependent variable. Similar rules can be incorporated with any of the distance measures previously suggested.

8.3 INTERPRETATION AND USES OF THE AID RESULTS

A hierarchical tree of subgroups as illustrated above in Figure 8.1 is the end result of a binary segmentation process. But in exploring patterns of relationships of predictor variables with the dependent ones, various types of intermediate result in the tree formation process will prove just as important. In AID II, for each subgroup the following information is available: distributions of individuals over the categories of each predictor; profiles of dependent variable means and standard deviations over the categories of each predictor; the BSS and η^2 (i.e. BSS/TSS) values for contiguous splits on each predictor; and the values of BSS and η^2 for the best binary split of the subgroup.

8.3.1 Uses of AID

Once the tree and intermediate output have been produced there are no hard and fast rules to guide the researcher in interpreting them. Indeed, the uses to which the survey analyst puts the results are varied. There are, however, four broad but not necessarily distinct types of application.

Firstly, binary segmentation has been proposed as a systematic way of unravelling complex variable dependencies. It is hoped that a careful examination of the output will enable the researcher to get a 'feel' for the data structure. Yet each data set will have its own pecularities, and often results will depend on such diverse things as how the researcher has defined his dependent variable and the form and number of categories of the predictors. However, the AID II output provides information which helps in identifying the data structure. The relative importance of various predictors at each stage of the tree can be assessed. Frequency distributions of predictor categories compared across subgroups are indicative of associations between predictors. The shape of the tree and the final subgroups it produces are suggestive of predictor effects and interactions between them. Of further importance here are the profiles of dependent variable means compared across subgroups. Yet at the end of the day there is often no substitute for intuition and common sense in such exploration.

An examination and interpretation of the results of binary segmentation may in some cases be a preliminary to a second type of approach to analysis of survey data. The aim may be to establish an economical prediction equation explaining a large part of the variation in the dependent variable. This aim is not unlike conventional regression analysis but it does allow categorized predictor variables and explicitly provides for potential interactions not specified in advance. The originators of the AID algorithm proposed that it be used in this way.

A third and very specific use of the procedure is to identify homogeneous, and often deviant groups. Here a minimal amount of information is required and the objective is clear at the outset. There is no real interest in shapes of effect, degrees of interaction, or patterns of explanation. It is simply an attempt to establish what combination of predictor characteristics is likely to identify individuals with extreme values on the dependent variable. A good example of this is found in Carlson (1968) where an AID analysis was used with success in establishing which combination of characteristics serves to identify potential problem drivers.

A further useful application of binary segmentation as a data-reduction device has been suggested by Lindsey (1973). Here a dependent variable takes on a large number of values and it is desired to group these according to and reflecting similarities between individuals on a number of other variables. The end result of a binary segmentation procedure facilitates this.

8.3.2 An example

The interpretation of AID results, their various applications, problems arising from the nature of the algorithm, and the recognition of complex features of particular data sets will be discussed using an example which is illustrated in Figure 8.2. Full details of the study are to be found in Silverman (1971). The dependent variable was price per 100 metres square at 1970 prices for a series of sales of flats in Edinburgh. The five predictors were categorized: floor in tenement (4 categories); daylight and sunlight standard (8 classes); location (13 categories); plan type (5 categories); size class (6 categories). Detailed definitions of these variables

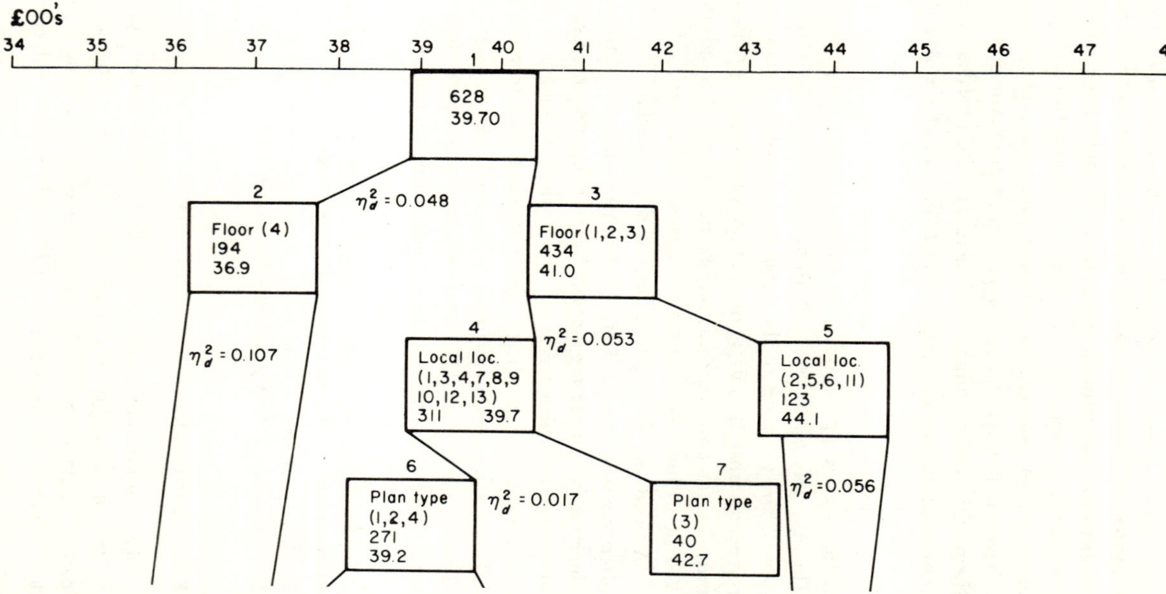

£00's
34
35
36
37
38
39
40
41
42
43
44
45
46
47
48
1
628
39.70
$\eta_d^2 = 0.048$
2
Floor (4)
194
36.9
3
Floor (1,2,3)
434
41.0
$\eta_d^2 = 0.107$
4
Local loc.
(1,3,4,7,8,9
10,12,13)
311 39.7
$\eta_d^2 = 0.053$
5
Local loc.
(2,5,6,11)
123
44.1
6
Plan type
(1,2,4)
271
39.2
$\eta_d^2 = 0.017$
7
Plan type
(3)
40
42.7
$\eta_d^2 = 0.056$

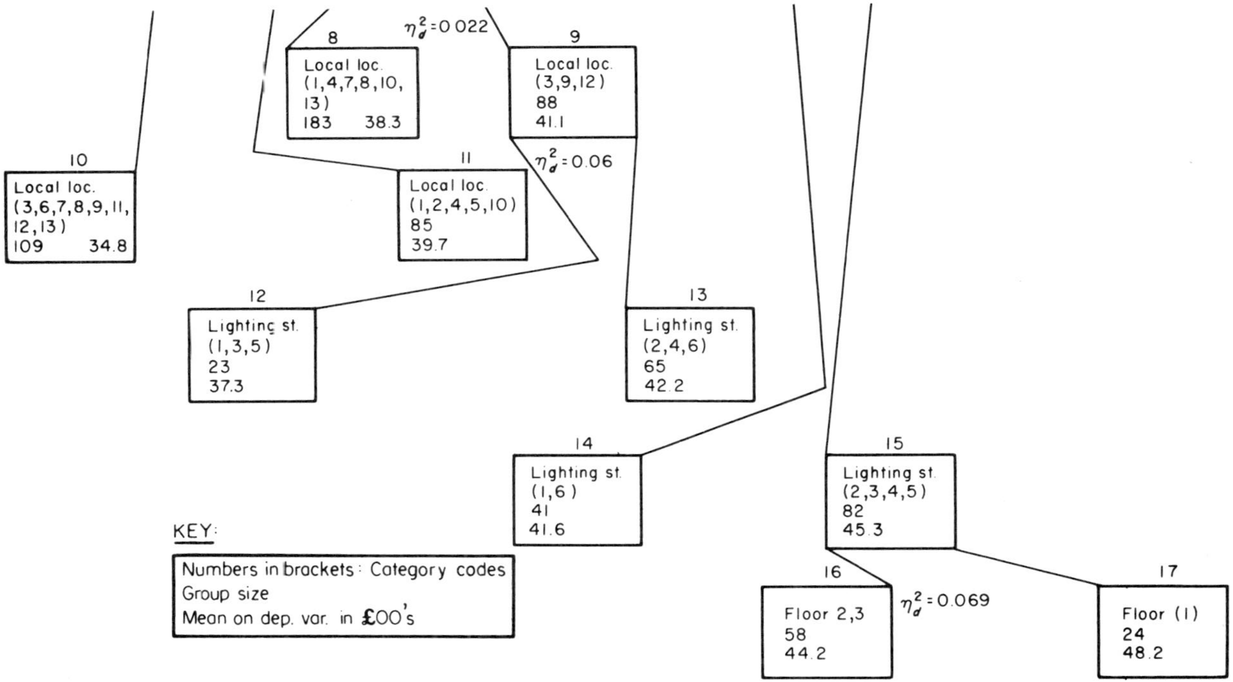

Figure 8.2. (After Silverman, 1971).

and some background to the study can be found in Appendix 1. The sample size of 628 is not large enough for any definite conclusions to be drawn, and it was acknowledged that there was a need for further research to test and refine the various hypotheses suggested by the AID runs. The conventional stopping rules were adopted with a minimum subgroup size of 25.

8.3.3 Detection of interaction

How can trees like Figure 8.2 with auxiliary results be interpreted and is their interpretation meaningful? Clearly with complex data these are not easy questions to answer and only rough guidelines can be suggested. An obvious feature of this tree is its asymmetry. This suggests some element of interaction between the predictors, though its nature is unclear from the diagram alone. Experimentation has shown that if sets of predictors are additive and non-interacting in their effects then a symmetric tree structure would be expected to emerge. In Orr's study of factors affecting various types of educational performance, several types of additive tree structures are illustrated (Orr, 1972). The relationship between symmetry and additivity may be made clearer by considering two simple sets of data. Table 8.1(a) examines the effect on a dependent variable of two dichotomous predictors, non-interacting and not associated. The first split would be made on the row dichotomy and within each row the effect of columns is the same, with a BSS of 100 in both subgroups. If in a more complicated but similarly structured example splits were made on the columns they would be expected to be the same in both subgroups. This is to be contrasted with Table 8.1(b) where although the predictor variables are also not associated the interaction effect is fairly strong. Again a row split might possibly be made but although the BSS for columns within the second row is increased to 400 it is driven to zero in the first row. Thus, even in the presence of other predictors, a split at some stage might be expected between columns within the second row. This could never happen within the first row. These are contrived, artificial and perhaps extreme examples but they give some justification for the intuitive assertion.

A symmetric tree structure is not of itself sufficient justification for the assertion of additivity of predictor effects. In addition the shapes of the profiles of dependent variables for categories of a predictor should be similar within various branches of a tree. The explanatory power of each predictor should also be similar at each stage in the symmetric tree structure leading to a situation where induced binary splits are much the same across all branches of the tree. Clearly it is possible for there to be an element of additivity in some branches of the tree but not others. In any additive situation the order of appearance of predictors, as in stepwise regression, is indicative of their importance in explaining the dependent variable.

The asymmetric nature of Figure 8.2 intuitively suggests interaction patterns. A more detailed examination of the intermediate output is now required to specify these more clearly. It is possible also that high degrees of association between the various predictors may be responsible for many features of the tree. Tables 8.2 and

Table 8.1(a).

		Predictor variable A		
		Category 1	Category 2	
Predictor variable B	Category 1	50,18,10,500	50,20,10,500	100,19,11,1100
	Category 2	50,15,10,500	50,17,10,500	100,16,11,1100
		100,16.5,12.25,1225	100,18.5,12.25,1225	200,17.5,13.25,2650

For overall group BSS for rows = 450
BSS for columns = 200
BSS for columns within row 1 = 100
BSS for columns within row 2 = 100

In each group of four numbers in the table, the first is the size of a group, the second its mean on a dependent variable, the third the standard deviation, and the fourth the sum of squares for that group. The BSS quantities can be calculated from these.

Table 8.1(b).

		Predictor variable A		
		Category 1	Category 2	
Predictor variable B	Category 1	50,19,10,500	50,19,10,500	100,19,10,1000
	Category 2	50,14,10,500	50,18,10,500	100,16,14,1400
		100,16.5,16.25,1625	100,18.5,10.25,1025	100,17.5,14.25,2850

BSS for rows = 450
BSS for columns = 200
BSS for columns within row 1 = 0
BSS for columns within row 2 = 400

Table 8.2.

Subgroup 2			
Plan type category	Number	BSS	Mean
1	76	5891888	3476
3	42	4469760	3744
2	60	1688832	3850
4	14	308820	3993
5	2	0	4084
	194	Max BSS/TSS = 0.054	

Subgroup 3			
Plan type category	Number	BSS	Mean
5	3	222141	3820
4	23	496305	3978
2	136	1838064	4019
1	198	7963536	4051
3	74	0	4393
	434	Max BSS/TSS = 0.025	

Table 8.3.

Subgroup 2				Subgroup 3			
Location category	No.	BSS	Mean	Location category	No.	BSS	Mean
13	3	2046518	2874	8	35	6725952	3674
11	6	2806448	3285	4	48	9622352	3871
6	14	5280336	3305	13	10	10042720	3927
9	14	6740800	3418	10	41	12375168	3929
12	17	8925984	3432	7	38	14056864	3980
3	22	10107280	3559	12	33	15818208	3998
7	14	11151136	3580	1	43	16652848	4078
8	19	11624144	3674	3	24	16693360	4132
10	11	11119456	3784	9	39	17114384	4154
4	16	9725184	3860	6	29	16654912	4226
5	18	7741232	3927	2	42	9333088	4436
2	19	5686240	3976	5	35	4621712	4437
1	21	0	4185	11	17	0	4605
Total	194	Max BSS/TSS = 0.107		434		Max BSS/TSS = 0.053	

8.3 are an example of this intermediate output presented for each predictor in subgroups as they develop. Such tables can be examined for predictors across subgroups for evidence of additivity or otherwise, and for effects of predictor associations. Subgroups 2 and 3 from the tree are flats on floor 4, and those on floors 1, 2 and 3 respectively. The categories of each predictor are ordered

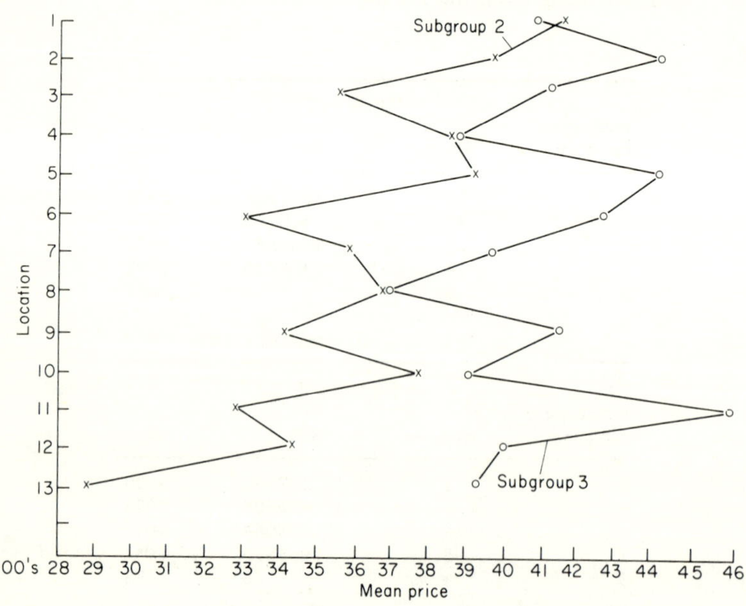

Figure 8.3.

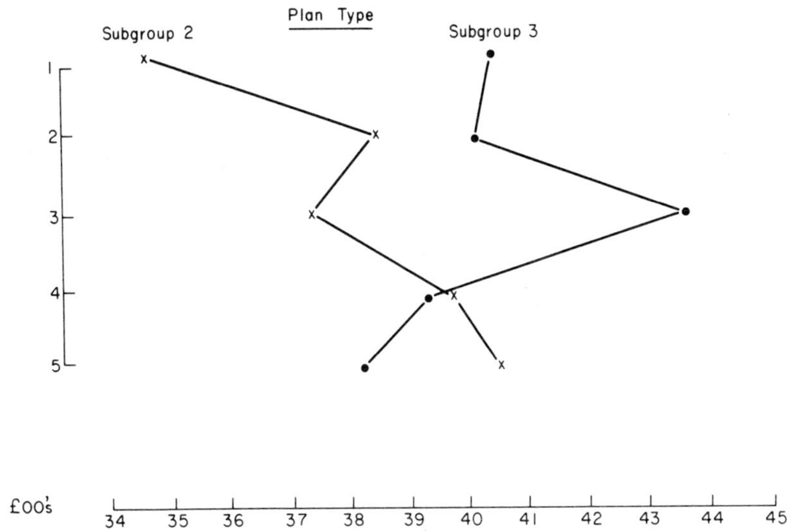

Figure 8.4.

according to their mean values on the dependent variable and the BSS for contiguous partitions are listed together with η^2 for the best split on that predictor. Another useful and illuminating way of displaying the profiles is to graph them as interaction diagrams. Figures 8.3 and 8.4 are the graphs for subgroups 2 and 3 on location and plan type. It is clear from both the tables and diagrams that the profiles of mean price differ considerably between the groups and consequently there is evidence of considerable interaction between both plan type, location and floor split. Additivity would be evidenced by roughly parallel lines in the diagrams. It may be further noted that when subgroups 2 and 3 are split further, both are induced by location but by totally different patterns of categories. The values of η^2 are also different: 10.7 per cent against 5.3 per cent.

Interaction is a complex concept and no steadfast rules for interpretation can be laid down. Each data set must be examined on its own merit. The results of the algorithm have reduced the mass of information to manageable proportions. A careful reading of the accumulating literature on experimentation with AID and its interpretation in published applications is recommended to the potential user. Sonquist (1970) gives some extreme examples of interactive situations to which a real tree may, to a greater or lesser extent, roughly conform. The most extreme form of interactive data will produce a completely asymmetric tree as in Figure 8.5. Here only certain combinations of categories of a series of predictors will produce a high (or low) mean value of the dependent variable and individuals not possessing this particular combination will have roughly the same mean. It is not sufficient for an individual to belong to a category of a particular predictor with a high mean effect. Lack of membership of certain categories of any of the other predictors is sufficient to counteract this and bring the dependent variable down to a common

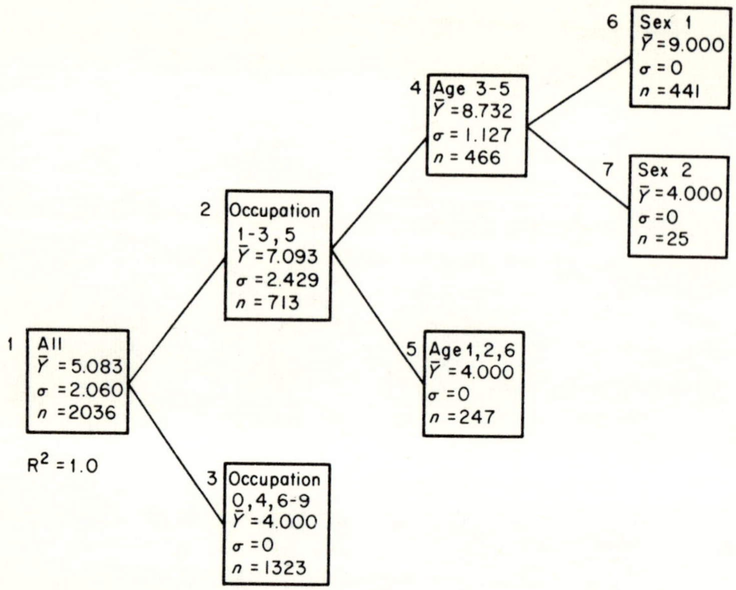

Figure 8.5. (Reproduced with permission from J. A. Sonquist, *Multivariate Model Building,* p. 156, Institute for Social Research, University of Michigan, Ann Arbor, 1970.)

low level. Positive effects are not additive over predictors. From experimentation with this and other types of interaction model Sonquist lays down some criteria for the detection of interaction patterns. One pointer to real interaction effects is a situation where there is a substantial increase in predictive power of a predictor when a subgroup is split off from a parent. This can be seen in the simple example of Table 1(b). In Silverman's study, in parent group 1 the best η^2 for location was 4.2 per cent. When the group had split on floor, it was 10.7 per cent and 5.3 per cent within subgroups 2 and 3 respectively. It has already been established from other considerations that location interacts with floor. If a predictor is highly associated with ones already used then this increase in predictive efficiency is further evidence of interaction, since under conditions of suppression and additivity its effectiveness should decline. The explanatory power of lighting standard declined considerably within groups 2 and 3 and other evidence from the output indicated no real interaction with the floor split. The fall in predictive efficiency of lighting standard in group 2 may of course be due to its high overall variation and association with floor. This can hardly explain the decrease in group 3 however where lighting standard within floors 1 – 3 was relatively heterogeneous. However, this is a point that must be fully explored before assertions of additivity or interaction are made. Further checks must be made on the increase in predictive efficiency in the two subgroups before they can be interpreted as indications of interaction effects. The increases must not be symmetric, i.e. they should either be quite high in one group and very low or even negative in the other group, or the profile of means should be in roughly opposite directions, since with correlated predictors, such results may emerge even with additivity.

8.3.4 Association among predictors

There are additional problems of interpretation of results which arise when predictors are associated in some way. In multiple regression analysis these are referred to as multicollinearities. Clearly when two or more predictors are highly associated, they are to some extent 'proxies' for each other. If one predictor is used to induce a binary split of a subgroup, particularly at early stages in a tree, then clearly it may be using up some of the explanatory power of the others. This is perhaps more important with additive effects, since with substantial interactions it is likely that these 'proxy' variables will appear later in some branches of the tree. However, it is possible that predictors which have large main effects might 'lose' the 'competition' for splitting, at some stage, to predictors with which they are closely associated. In such circumstances they may only turn up at very late stages in the tree with reduced predictive power and often may not appear at all. In multiple regression associated predictors do not of course create problems since estimates of effects of each predictor are made simultaneously and net of the effects of others. In stepwise regression and its counterpart here, AID, close associations between predictors do have this 'supressing' tendency, and it is clearly necessary to be wary of the conclusion that such predictors are unimportant. If the sole purpose of the analysis is to define subgroups to maximize explained variation in a prediction model they may well be ignored, since their explanatory power will be subsumed in other predictors. However, if one is interested in how they are related to the dependent variable and how they interact with other predictor variables, it is clear that they must be examined more closely. The strategy is to use AID results to identify suppressed variables. A series of runs can be made, omitting in each subsequent run one or more of the identified highly associated variables. A more general possibility is to examine effects of predictors suppressed in one run by performing another run with the predictors used in the first few splits of the earlier run left out.

An examination of the subgroup predictor frequency distributions serves to identify associated predictor variables. These are printed as part of the output but comparisons across subgroups are facilitated if they are converted to percentage distributions. Tables 8.4 and 8.5 present such distributions for location and plan type within subgroups 2 and 3. From these and the absolute frequencies of Tables 2

Table 8.4. Percentage distribution (Location)

	1	2	3	4	5	6	7	8	9	10	11	12	13	
Subgroup 2	10.8	9.8	11.3	8.2	9.2	7.2	7.2	9.8	7.2	5.7	3.0	8.8	1.5	100% = 194
Subgroup 3	9.9	9.7	5.5	11.1	8.1	6.7	8.8	8.1	9.0	9.4	3.9	7.6	2.3	100% = 434

Table 8.5. Percentage distribution (Plan type)

	1	2	3	4	5	
Subgroup 2 (Floor 4)	39.2	30.9	21.6	7.2	1.1	194 = 100%
Subgroup 3 (Floor 1, 2, 3)	45.6	31.3	17.0	5.3	0.7	434 = 100%

and 3 it can be seen that both location and plan type are relatively independent of subgroup. Floor 4 is similar to floors 1, 2 and 3 combined in the distribution of cases over location and plan type categories. Neither is being suppressed by the floor split of the first parent group.

Close associations between predictors within subgroups is often evidenced by their being closely competing candidates for the optimum split of that subgroup. This can be recognized at the outset. For the parent group 1 in the present example the best split was on floor with $\eta^2 = 0.048$, against 0.044 for lighting standard, and 0.042 for location. It has been noted that location is relatively independent of the floor split and it is clear that suppressor effects are not operating here. Indeed location is next used in both branches as the best binary split predictor with recognizable interaction effects. However, it was known at the outset that a high natural lighting standard (usually associated with flats on upper floors) is related closely to a low standard of stair access and *vice versa.* This association is evidenced by Table 8.6 extracted from the AID output, and expressed in percentage form. Here practically all floor 4 cases are concentrated in two of the lighting standard categories, whereas the distribution is more evenly spread for floors 1, 2, 3 combined. Lighting standard is perhaps being suppressed but it does play a part in later stages of one branch of the tree when it induced further splits of floors 1, 2, 3.

Table 8.6. Lighting standard

	Floors 1–3 Subgroup 3		Floor 4 Subgroup 2	
Category	Number	%	Number	%
1	127	29.2	0	0.0
2	51	11.7	9	4.6
3	74	17.0	2	1.0
4	32	7.4	11	5.6
5	92	21.2	88	45.4
6	53	12.2	83	42.9
7–8	5	1.2	1	0.5
Total	434		194	

8.3.5 Data reduction and predictive models

Data-dredging begins with the results of AID which present a suitable set of summary information and a few guidelines. The main effect of each predictor on the dependent variable can be examined from the summary table of means and standard deviations for the overall group. This will also indicate the important variables. We can see in the example that the predictor 'size class' has no real effect, and since it never appears in the tree it can be discounted as an explanatory factor either on its own or interacting with others. The order of introduction of predictors gives an indication of their relative importance although care must be exercised here in the presence of associated predictors and when certain predictors are important in some branches of the tree but not in others. Thus plan type has marginal overall importance but plays a considerable part in discriminating prices in subgroup 4. The results also indicate another aspect of data structure, i.e. partial association. How is one predictor related to the dependent variable controlling for other predictors? This can readily be answered by examining the subgroup mean profiles and these may give some justification for the assertion of interaction if the type and strength of association differs for different combinations of categories of other predictors defining the various subgroups.

However, as has been previously mentioned, there is often a specific end to data dredging of this nature. The aim may be to establish a predictive model; the development of an equation with the aim of maximizing the explained variation in the dependent variable consistent with some degree of simplicity and economy. It is clear that this model will be more useful if it reflects the types and patterns of effects that the hypothesis generation procedure has suggested. Linear models where the effects of predictors are assumed to be additive have been the traditional approach to such problems. With interval scale predictor variables multiple regression models are fitted and multicollinearities of varying degrees are dealt with implicitly. The estimation of effects with nominal predictors proceeds from the adaptation of the multiple regression model to non-orthogonal predictors by the dummy variable regression methods of Yates (1934), and programmed as Multiple Classification Analysis (MCA) by Andrews, Morgan and Sonquist (1967). The simplest model to assume is additive in the predictors with a series of dummy variables for each predictor. This can be tested using a regression program and some measure of the level of variation explained gives an idea of its adequacy. The use of AID here is to suggest interaction terms. Then improved prediction models may be established using new variables which are compound categories of the predictors reflecting suggested interactions. Thus AID used as an exploratory device suggests plausible models which regression and other techniques can help to support.

The formulation of such models should have an equation as simple as is consistent with explaining a large part of dependent variable variation and should include only the important predictors detected by AID unless of course theory suggests the inclusion of other predictors. Thus the example of Figure 8.2 above resulted in eight final subgroups and a successful prediction equation is given by eight effects corresponding to membership of the groups in terms of the combinations of categories of the various predictors they represent. It is possible that a simpler model incorporating some predictors on their own and others in combination suggested by first order interaction patterns may, when fitted by regression, give almost as good a fit in terms of explained variation. In this case the latter approach may give more interpretable results. In the example, a regression model using plan type and lighting standard as a separate series of dummy variables and a 15 point categorization of various combinations of the location and floor variables as suggested by the output of a series of AID runs gave a very powerful predictive model. Separate models were established for subgroups 2 and 3 and these were even more informative. Sonquist (1969) suggests a strategy for using AID and dummy variable regression together to establish models which he refers to as a kind of 'middle-range causal proposition'. He emphasizes the need for separate analysis of subgroups which are suspected of having a completely different pattern of effects. Successive runs of AID with different deletions of variables successively refine the plausible interaction terms. Sonquist recommends that in certain circumstances much can be achieved by performing an AID run with a new set of predictors on residuals from an earlier run. This may be desirable, for instance, where there are two sets of predictors logically distinct in a postulated causal chain.

8.3.6 Conclusion

There are many published examples of the application of AID in survey analysis. Many of these have taken advantage of the flexibility of the algorithm, using it as a powerful search and hypothesis generation procedure. In others only the AID tree diagram has been published and a few sentences of interpretation given. In these cases one can see many of the dangers which the unthinking, uncritical and unimaginative application of heuristic algorithms produces. Often another reader might reach a totally different impression of the structure of effects. The trees are presented in isolation and given as final evidence for the conclusions reached. This is not a useful way to proceed; rather, many of these authors should have given more thought to the model building objectives. Such examples serve only to discredit the development of flexible heuristic approaches. Many of the examples are based on samples so small (70 in two of the applications) as to preclude even two way tables of means, to say nothing of an examination of the effects of up to ten predictors. Some applications are given in the references.

Whether binary segmentation answers all the questions desired of an exploratory approach to data analysis remains, as yet, an open question. Continued experimentation with and development of such approaches will provide evidence in this direction. There are still many unsolved problems in determining whether they do in fact what they purport to do. Some of these will be discussed in Section 8.5.

8.4 THE AID III VERSION: FURTHER MODIFICATIONS AND FLEXIBILITY

The earliest version of binary segmentation for one interval scale dependent variable using the AID between group sum of squares criterion embodied the simple features outlined previously. However, early experimentation in its use suggested ways in which it might be refined by introducing further features which could improve the search and interpretation process. These are available in the computer algorithm known as AID III (Sonquist, Baker and Morgan, 1971).

Changing variable definitions

It has been noted that the categorization of the predictors may affect the operation of the segmentation. Recoding facilities exist in the new version which enable the user to try different ways of defining categories. One can also use these to rearrange categories for 'monotone' splits, generate new variables, transform existing variables, build indices, or define new predictors in terms of combinations of more than one existing predictor. A facility is available which allows exclusion of cases according to almost any specification. Truncation of the dependent variable is also possible. These facilities enable the user to experiment with different variable specifications and to examine their effects on the segmentation.

Forcing splits

Apart from the simple restrictions imposed by declaring certain predictors 'monotone', the early version of AID gave a branching process that was unconstrained. There are several situations where it may be desirable for the analyst to impose restrictions on the process. These have the effect of further limiting permissible splits. The simplest option available is to start the analysis not from the overall group but from a predefined set of subgroups. One can *a priori* force the first few splits to give different subgroups in which there is special interest, usually because it is expected that the pattern of effects of the various predictors will be totally different in each group. Forcing splits has other uses, particularly in replication studies. One may take the tree produced from one set of data, or a subsample, and force it on another set to compare the results. This facility also enables the analyst to examine the effects of using a particular predictor at a stage other than the one which may have resulted from an unconstrained process, or to ensure that a particular predictor is used early in the splitting process. The latter option is particularly useful if this predictor is logically prior to others.

Ranking predictors

Further restrictions are embodied in the predictor ranking options of AID III. Predictors are assigned a rank which determines the order in which they will be used in the branching process. Rank zero has a special use, since all predictors that have this rank are never used in the splitting process but statistics and other intermediate information are computed for them. These are usually variables which may not have a logical place in the development of a predictive model, but may have interesting, if peripheral, relationships to the dependent variable. They may be effects of the dependent variable and not causes of it.

In starting the operation of the algorithm only predictors with the highest rank are considered. It is only if none of these meet the split eligibility criteria that predictors with the next highest rank are considered. This continues until a rank is reached at which a partition can be made. In various branches of the tree, predictors with ranks that have been abandoned are not considered again, although the summary tables and their profiles of dependent variable means are provided in the output. Various branches of the tree may be considering different rank levels at any stage of the segmentation process. A somewhat less restrictive option referred to as the range-ranking option allows partitioning on predictors whose ranks are in the same specified range as the rank which is currently under consideration at some stage in the tree. The purpose of these options is to provide the user with facilities permitting him to use his theoretical knowledge of a rough set of causal priorities in controlling the order of appearance of predictors in the tree. Background factors such as area construct variables are considered to be 'early' in their effects, and it may be desirable to examine these first. This approach is also useful if it is felt that there are likely to be interactions between variables at different levels, since controlling the tree in this way enables these to emerge more clearly. The current ranking options allow up to ten rank levels.

Forcing symmetry

A further restriction on the splitting process, enabling the analyst to examine the effects of the same predictor for two subgroups in a branch of the tree, is provided by the option of forcing symmetry. If one of a pair of groups in the branch of a tree has been split on particular categories of a predictor, the other is split in the same way. To prevent substantial loss of predictive power a comparison is made between the best split for this group and the symmetric split. One can set a premium on symmetry by specifying that the symmetric split will only be made if its predictive power, as measured by the between sum of squares for resulting subgroups, is at least as high as some percentage of the power of the best split (90 per cent for example). The overall tree will thus be only partially symmetric since this premium on symmetry will operate to allow the symmetric restriction only in certain branches. The aim of forcing symmetry is to see if an additive model is an attractive proposition.

The lookahead option

One problem arising from hierarchical splitting processes in general and binary segmentation in particular is that at the conclusion of the analysis subgroups have obviously been defined and developed purely on a sequential basis. The process operates to maximize the explanation at each stage and there is no guarantee that this will produce the best k-group partition. Inefficient early binary splits may not lead to the best overall set of subgroups. Any direct attempt to find the best k-group partition directly will be computationally infeasible since it would involve considering all possible k-group partitions. This is, indeed, the main reason for using a sequential approach. However, although not attempting to find directly a best global partition, the AID III 'lookahead' option does attempt to improve on the stage by stage optimization. This operates to find a binary split which although not necessarily optimally separating subgroups at a given stage, achieves optimum explanatory power over subgroups formed by the series of binary splits following it. For the group under consideration at any stage the split induced by optimizing for each of the predictors is examined. For m predictors there will be m tentative splits under consideration. The best splits for each predictor on these subgroups is then obtained. One now has m^2 possible two-stage trees under consideration. This process could be continued for r stages with m^r possible trees. In principle one could make the split at the first stage which produced the largest between group sum of squares for the final groups. Clearly this lookahead option could involve a tremendous amount of computation and information storage capacity were it not restricted. The current version of AID III limits the lookahead steps to three, including the first split. With the zero option lookahead is not invoked and the group is split on the best predictor for a single stage. Whichever option is selected, it is used at all levels of the tree. If the lookahead option is used a further stopping rule is invoked. The minimum level of explanatory power for a sequence of steps must be higher than that acceptable for a single step. A minimum for this can be set by the user.

Segmentation imposed by comparison of regression on a covariate

A final feature of AID III is the introduction of criteria which form partitions of subgroups, not solely in terms of maximal separation on the dependent variable, but by maximally different relationships between the dependent variable and a control variable or covariate. Analysis of covariance is common enough in the traditional analysis of experimental data. It is used either to adjust a dependent variable before the main effects of the experimental factors are examined, or to examine differences in the regressions of the dependent variables on the covariate between these factors. With survey data similar questions arise. We may, for example, wish to investigate how the regression of earnings on level of education varies in different occupation, area, and background contexts. We wish to define subgroups whose regressions differ as much as possible from the regression in the parent group. The criterion used maximizes the difference between the total sum of squares explained by separate regressions in the subgroups and the overall sum of squares explained by the regression in the parent group, i.e.

$$\sum_{j=1}^{2} n_j(\bar{y}_j - \bar{y})^2 + \left[\sum_{j=1}^{2} \left[\frac{\left(\sum_{i=1}^{n_j} (y_{ji} - \bar{y}_j)(Z_{ji} - \bar{Z}_j) \right)^2}{\sum_{i=1}^{n_j} (Z_{ji} - \bar{Z}_j)} \right] - \frac{\left(\sum_{1}^{n} (Y - \bar{Y})(Z - \bar{Z}) \right)^2}{\sum_{1}^{n} (Z - \bar{Z})^2} \right]$$

where Z is the covariate (see Cochran and Cox, 1957). It will be seen that in the above formula the first term is essentially the sum of squares due to differences between the subgroup means of the dependent variable and would have been the criterion in an ordinary binary segmentation. However, the second term is that contributed by differences in the regression slopes in the subgroups. It would be possible for the first term to swamp the second if the two groups had large differences in mean levels of y and even if the regression slopes were not very different for two subgroups a split might be induced. In examining differences in explanatory power of the overall regressions the mean levels are as important as the slopes. However, it may be desirable in some circumstances to isolate subgroups solely in terms of whether their regression slopes are widely different. AID III incorporates such a slopes-only analysis. To adjust for differing mean levels of y for categories of a predictor, weighted slopes over the k classes of a predictor in any group are found with sum of squares explained by the weighted slopes.

$$\left(\sum_{i=1}^{k}\sum_{j=1}^{n_i}(Y_{ij}-\bar{Y}_i)(Z_{ij}-\bar{Z}_i)\right)^2 \Big/ \sum_{i=1}^{k}\sum_{j=1}^{n_i}(Z_{ij}-\bar{Z}_i)^2 .$$

where n_i is the number of observations on category i of the predictor. The difference between the sum of two such quantities for subgroups induced by a binary split on a predictor and one for the parent group, is the criterion to be maximized for that predictor. Optimal binary splitting occurs when these quantities are then optimized over all predictors. Giving covariates the rank zero may be used to examine further their relationships with the dependent variable. Although the covariates are not considered as predictors for the purposes of inducing a split, the output may uncover non-additivities in subgroups which are concealed in the regression analysis for the whole sample.

Conclusion

The wide range of flexible options available with AID III helps the analyst further in the examination of the pattern of relationships. Again the researcher's strategy should include successive refinement of his predictive model by trials of various options. As yet many of these new options have not been widely used although clearly they have great potential. Whether the extra computer effort involved is made worthwhile by the flexibility provided for the user is still an open question (see Muxworthy, 1972).

8.5 THE BEHAVIOUR OF AID: OBJECTIVITY AND STABILITY

Any heuristic algorithm for exploring data structure, if it is to be useful, must have interpretable results. The uses of a particular method of binary segmentation, AID, have been examined in some detail. Binary segmentation is not the only semi-automatic procedure available but enough has been said to indicate that it will in general be a more powerful approach than less directed searches for structure. In any event, its aims are limited enough for one to believe that it will work successfully as a hypothesis generation procedure. The tree diagram is an intuitively appealing way of summarizing many of the salient aspects of the data structure. The potential for producing a large amount of summarized intermediate information forms a systematic and directed data analysis strategy. It is model-free, so that its use in formulating models for subsequent testing is valuable.

Types of data structure are so diverse that the procedure may work very well in some applications but not so well in others. The combination of 'distance criterion' adopted, type and level of data, nature of stopping rules, and the actual structure of variable relationships in a given data set determines the behaviour of the algorithm. Investigation of this behaviour must proceed by continued experimentation both on real data sets and simulated artificial data structures. It is to be expected that accumulating evidence will demonstrate how and where it may be effectively used, although such evidence may also point to situations where interpretation of its results may be difficult. The establishment of adequate distribution theory for the criteria used would lead to a more objective evaluation; but this is likely to continue to be an intractable problem. Yet even without such developments if its use is restricted to theory building and hypothesis generation its power as a data analysis tool is evident.

In this context it is unfortunate that many of the published applications, in particular of AID, have not proceeded in an exploratory mode. Research conclusions have often been based on a single tree with supplementary output ignored. No attempt has been made in these examples to work towards a model for the data structure in the ways that have been suggested. The dangers of this are obvious and Cramer (1971) was highly critical of AID in his review of the algorithm. However, he bases this criticism on comparisons with conventional analysis of variance. The use of AID as an exploratory device rather than as a framework for supportive tests is largely ignored. A careful examination of all the features of the results are required before plausible structures and models are hypothesized. To this extent it is not fully automatic.

It may be recognized that binary segmentation, and in particular AID, may not always perform well for certain types of structure. However, it would be useful if it did satisfy two desiderata. First, in any application to a particular set of data different researchers should reach roughly the same conclusions. Second, the technique should also have reasonable sampling stability. Similar results should emerge from an application of the algorithm to independent random samples from some multivariate population structure.

Sonquist and Morgan (1964) and Sonquist (1970) have accumulated much evidence of the behaviour of AID on contrived data sets. An important feature observed was a tendency of the algorithm to split off a large number of small groups based on many predictors if the distribution of the dependent variable was highly skewed within subgroups. Such results became difficult to interpret. A suitable transformation of the dependent variable, as in common elsewhere, may help. It also appears that the algorithm may not work well if some of the important predictors show a marked departure from a uniform distribution over categories. Similar remarks apply to predictors constrained to be monotonically related to the dependent variable if they are asymmetrically distributed. This is largely due to the nature of the criterion. Skewed predictors tend to be ignored in favour of symmetrical and uniform ones, even if it is evident from intermediate output that over all categories they are more closely related to the dependent variable. With predefined categorizations little can be done here, except for some *ad hoc* regrouping. However, for interval level predictor variables a careful choice of class limits can help. Closely related to this problem is the experimental observation that many predictors induce binary splits which result in subgroups of roughly equal size but with not necessarily widely different dependent variable means. The criterion also appears sensitive to the number of categories of predictors. This was not explicitly considered by Sonquist's experiments but other evidence indicates that predictors with larger number of categories will tend to be favoured for the binary split even though they have overall explanatory power similar to others (Fielding and Sheperd, 1973; Fielding and O'Muircheartaigh, 1976).

AID has been shown to recover the original data structure quite well in experimental work on both additive and extreme forms of interactive structures. It may be conjectured that it would also behave quite well for less extreme but reasonably well structured data sets (Sonquist, 1970). Sonquist and Morgan (1964) also examined the effect of introducing spurious predictors and random 'noise' in their models. The 'noise' or error in a model can be treated as variation within the groups formed by a full configuration of predictor category combinations. It is clear that the real effect of the size of this noise is related to how the stopping rules are set. The number of subgroups eligible for splitting at any stage increases with error since subgroups will contain a larger within group sum of squares. The presence of error also decreases the number of groups capable of being split since each split transfers a smaller proportion of total sum of squares to between group sum of squares. More crucial is the experimental observation that the algorithm tends to produce spurious splits in the presence of considerable noise, often early enough in the structure to cause concern. If a test of significance could be devised and stopping rules modified accordingly, this might be less of a problem. Since the criterion for splitting is to choose the maximum of a series of two-group BSS, this may be quite high by chance, even if a spurious predictor is under consideration. Also predictors with small but substantively insignificant effects might be allowed to operate for the same reason. This is important in view of the fact that large residual error and low explained variation are the rule rather than the exception in much social research. Unless there is development of an adequate statistical theory, modifications of stopping rules can only be suggested by continued experimental work.

The empirical investigations have helped to aid the researcher in his interpretation of results, and will continue to do so. However, they do not fully address themselves to the problem of generalization from samples. There is a continuing need for empirical investigations of the sampling stability of binary segmentation, bolstered where possible by sampling theory. Fielding and Sheperd (1973) have reported findings on comparisons of random subsamples of various sizes from a large data set. Without suitable summary statistics these comparisons must often be made by inspection. Comparison of the tree structures must usually be made visually. It is noted in these investigations that often the profiles and order of inclusion of variables differ considerably between subgroups for subsamples as large as 3,000. This is larger than most of those used in the applications quoted. However, there may be many paths in a tree which lead to the same end, and a comparison of the final subgroups may be a more appropriate basis for an evaluation of stability. It is reasoned that if these are similar then a careful examination of the data would lead to the same postulated model in any case. Yet even here considerable instability was noted. Although broadly comparable the final groups for runs on different subsamples of the same size produced many striking discrepancies. The conclusion reached is that very large samples are required if generalization from a single run is not to be misleading.

Comparisons were also made of the overall explained proportion of variation, i.e. the ratio of the between group sum of squares for final subgroups to the total sum of squares. This was apportioned to the various predictors as they were introduced. Marked differences were to be observed even for large samples. Intermediate output plays a large part in careful interpretation and consequently these were compared across subsamples for similar subgroups. Sampling instability was smaller here and the profile of dependent variable means for each predictor was stable as is to be expected. However, due to the nature of the BSS criterion, and especially in the presence of closely associated predictors, splits on different predictors were obtained at a particular stage in the tree in different subsamples.

It may be noted that a likelihood ratio test statistic of the null hypothesis that the categories of a predictor have equal dependent variable means against the alternative that the category means fall into two distinct groups with equal means within each has been developed (Scott and Knott, 1974). The criterion maximum BSS/TSS for one predictor can be shown to be proportional to this, so at least it has some basis in conventional statistical methodology. Scott and Knott simulated the null distribution for categories of equal size. From an examination of their critical values it appears that the stopping rules conventionally adopted for the size of BSS/TSS may well be much too low. Clearly for a full evaluation, a stopping rule based on the complicated distribution of the criterion for unequal category sizes, across predictors, and conditional on the stage in the algorithm reached would be required. However, the limited evidence points to the desirability of setting the minimum BSS/TSS much higher than is conventionally adopted.

Recent work by Hartigan (1972) is also relevant to these distributional problems. He discusses a simultaneous clustering scheme of an $n \times k$ respondent-variable matrix into blocks. The algorithm uses the maximum BSS criterion but it is applied to the simpler case of two orthogonal predictors with one observation per category combination. He finds the asymptotic null distribution of the criterion for this simple case. Though the usual data structure to which AID analysis is applied is much more complicated some of his results provide conjectural explanations of certain features of the behaviour of AID. He shows, for instance, that in a null situation the criterion will almost certainly result in splitting into subgroups of roughly equal size. In this case, the resulting subgroups for each predictor will contain $m/2$ cases, where m is the number of categories for the parent group under consideration. A similar phenomenon may well carry over into a situation with unequal category sizes. Here predictors would induce splits with resulting subgroups as nearly equal as possible. It is a reasonable conjecture that this may also happen in the non-null case where a predictor's category mean differences were roughly equal. Certainly this is a feature of AID that has frequently been noted in empirical investigation. Hartigan develops the normal asymptotic null distribution from this fact. He shows that the expected values of the criterion for a predictor are proportional to its number of categories. This is also evident from the small sample simulations of Scott and Knott (1974). This is in any case an intuitive result. One would expect the maximum of a number of similarly distributed quantities to be directly proportional to that number. Similar results are to be expected in the non-null case since here the criterion would be higher for a predictor having a larger number of categories even if, in a general sense, its strength of association with the dependent variable was similar to other predictors. The theory is limited and cannot be directly extended to the complicated situations existing in practice: non-normal dependent variables, non-orthogonal associated predictors, and predictors not uniformly distributed over categories. However, such considerations suggest a transformation of the criterion based on the number of categories before a comparison across predictors is made. In Hartigan's simple scheme division by the number of categories is suggested. However, modifications for more complicated situations have not yet been developed.

Hartigan's work also suggests a stopping rule for the segmentation process which may be preferable to existing ones. At present the AID algorithm considers stopping the splitting process for each subgroup in isolation unless the specified maximum number of final subgroups has been reached. It has been noted above that at any stage in the development of the AID tree, the algorithm chooses to operate on the subgroup with the largest potential for increase in predictive efficiency (as assessed by the within group sum of squares), rather than the one with the actual largest increase. Hartigan suggests a broader approach to the stopping process than the consideration of groups in isolation will allow. He considers the increase in predictive efficiency across subgroups relative to the unexplained variation to be a more effective criterion.

Experimentation with AID and other binary segmentation procedures and the establishment of an effective theory are in their early stages. Some directions for development have been noted. The work that has been done suggests that such an approach is not the 'philosopher's stone' of survey analysis. However, the simplicity of the criteria, and the ease with which the information can be transmitted in a tree diagram, together with the intuitive appeal of the procedures, provide potential for wider application. The emphasis throughout this chapter has been on caution. There are no quick and easy answers. With care, skill, informed judgement and an understanding of the true purpose of exploratory research, the user of binary segmentation procedures can expect them to be an effective tool in illuminating structures in multivariate data.

8.6 APPENDIX

Background and variables in Silverman's study

Study units. The study area was a well-defined collection of tenement flats in Bruntsfield, Edinburgh. The study unit was each sale of a flat in this area between January 1 1920 to November 1 1970. A total of 628 such sales was identified from the Register of Sasines.

Dependent variable. Price per 100 metres squared, and expressed at 1970 prices.

Predictors.

(1) *Floor:* the floor of the tenement on which the flat was located; 1,2,3,4.

(2) *Daylight/Sunlight Standard:* A compound categorization of various types of pass and failure on prescribed daylight and sunlight standards.

1. Fails daylight standard, fails sunlight standard in both living room and kitchen.
2. Fails daylight standard, fails sunlight standard in living room but passes in kitchen.
3. Fails daylight standard, fails sunlight standard in kitchen but passes in living room.
4. Passes daylight standard, but fails sunlight standard in both living room and kitchen.
5. Passes daylight standard, fails sunlight standard in living room but passes in kitchen.
6. Passes daylight standard, passes sunlight standard in living room but fails in kitchen.
7. Fails daylight standard, but passes sunlight standard in both living room and kitchen.
8. Passes all three standards.

(3) *Location:* The study area was divided into 13 cells.

(4) *Plan Type:* The study units were one of five distinct plan types.

(5) *Size Class:*

1. 50–59 metres square
2. 60–69 metres square
3. 70–79 metres square
4. 80–89 metres square
5. 90–99 metres square
6. 100–109 metres square

REFERENCES

Asterisked items report applications of binary segmentation which are not referred to in the text.

* Agostini, J. M. (1966), 'A method of market segmentation', *Advertiser's Weekly*, January 7th, 22–24.

Andrews, F. M. and Messenger, R. C. (1973), *Multivariate Nominal Scale Analysis*, Ann Arbor: Institute for Social Research, University of Michigan.

Andrews, F. M., Morgan, J. N. and Sonquist, J. A. (1967), *Multiple Classification Analysis*, Ann Arbor: Institute for Social Research, University of Michigan.

*Belson, W. A. (1956), 'A technique for studying the effects of a television broadcast', *Appl. Statist.*, **5**, 195–202.

Belson, W. A. (1959), 'Matching and prediction on the principle of classification', *Appl. Statist.*, **8**, 65–75.

Bouroche, J. M. and Tennenhaus, M. (1972), 'Some segmentation methods', *Metra*, **7**, 407–418.

Boyce, A. J. (1969), 'Mapping diversity: A comparative study of some numerical methods', in *Numerical Taxonomy*, A. J. Cole, ed.

*Bynner, J. (1973), 'AID versus discriminant analysis in a study of smoking', *Unpublished Mimeo*.

Carlson, W. L. (1968), 'Identifying the problem driver from state driver records', *Highway Safety Research Institute*, Michigan.

Cellard, J. C., Labbe, B. and Savitsky, G. (1967), 'Le programme Elisee presentation et application', *Metra*, **3**, 511–519.

Cochran, W. G., and Cox, G. (1957), *Experimental Designs* (2nd. Edn.), New York: Wiley.

Cormack, R. M. (1971), 'A review of classification', *J. Royal Statist. Soc. (A)*, **124**, 321–353.

Cramer, E. M. (1971), 'Review of Sonquist, J. A., Multivariate model building', *Psychometrika*, **36**, 440–442.

Engelman, L., and Hartigan, J. A. (1969), 'Percentage points for a test of clusters', *J. Amer. Statist. Ass.*, **69**, 1647–1648.

Edwards, A. W. F. and Cavalli-Sforza, L. (1965), 'A method for cluster analysis', *Biometrics*, **21**, 362–375.

Fielding, A., and Sheperd, J. W. (1973), 'The sampling stability of AID', contributed paper to *39th Session Int. Statist. Inst.*

Fielding, A., and O'Muircheartaigh, C. A. (1976), 'Binary segmentation in survey analysis, with particular reference to AID', to be published.

Fisher, R. A. (1936), 'The use of multiple measurements in taxonomic problems', *Annals of Eugenics*, **7**, 179–188.

Fisher, W. D. (1958), 'On grouping for maximum homogeneity', *J. Amer. Statist. Ass.*, **53**, 789–798.

Fleiss, J. L. (1973), *Statistical Methods for Rates and Proportions*, New York: Wiley.

*Forsyth, J. C., and Mercer, R. G. (1973), 'The emigration of graduates from Scotland', *Regional Studies*, **17**, 95–107.

Gillo, M. W. (1972), 'MAID: A Honeywell 600 program for an automatised survy analysis', *Behavioural Science*, **17**, 251–252.

Gillo, M. W., and Shelly, M. W. (1974), 'Predictive modelling of multivariable and multivariate data', *J. Amer. Statist. Ass.*, **69**, 646–653.

Goodman, L. A., and Kruskal, W. H. (1954), 'Measures of association for cross classification', *J. Amer. Statist. Ass.*, **49**, 732–763.

Gower, J. C. (1971), Contribution to discussion of Cormack, A review of classification, *J. R. Statist. Soc.*, **134**, 360–365.

Hartigan, J. A. (1972), 'Direct clustering of a data matrix', *J. Amer. Statist. Ass.*, **67**, 123–129.

Heald, J. I. (1972), 'The application of the automatic interaction detector programme and multiple regression techniques to assessment of store performance and site selection', *Oper. Res. Quart.*, **23**, 445–457.

*Himmelweit, H. T., and Swift, B. S. (1969), 'A model for understanding of school as a socialising agent', in *Trends and Issues in Developmental Psychology*, Mussen, P., Langer, J. and Covington, M., eds., London: Holt Rinehart & Winston.

*Hutchison, D., & Littlejohn, S. M. (1976), 'The impact of social science on flows from school to university', *Research in Education*, Forthcoming.

Johnson, R. L., and Wall, D. D. (1969), 'Cluster analysis of semantic differential data', *Educ. Psychol. Measur.*, **29**, 769–780.

Kendall, M. G., and Stuart, A. (1968), *The Advanced Theory of Statistics*, Vol. III, London: Griffin.

Lazarsfeld, P. F. (1955), 'The interpretation of statistical relations as a research operation', in *The*

Language of Social Research, P. F. Lazarsfeld and M. Robenberg, eds., New York, The Free Press.

Light, R. J., and Margolin, B. H. (1971), 'An analysis of variance for categorical data', *J. Amer. Statist. Ass.*, **66**, 534–544.

Lindsey, J. K. (1973), *Inferences from Sociological Survey Data*, Amsterdam, Elsevier.

McNaughton-Smith, P. (1963), 'The classification of individuals by the possession of attributes associated with a criterion', *Biometrics*, **19**, 364–366.

Messenger, R. C., and Mandell, L. M. (1972), 'A modal search technique for predictive nominal scale multivariate analysis', *J. Amer. Statist. Ass.*, **67**, 768–772.

Morgan, J. N., and Messenger, R. C. (1973). *THAID, A Sequential Analysis Program for the Analysis of Nominal Scale Dependent Variables*, Ann Arbor: Institute for Social Research, University of Michigan.

Morgan, J. N., and Sonquist, J. A. (1963), 'Problems in the analysis of survey data', *J. Amer. Statist. Ass.*, **58**, 415–434.

*Morrison, D. G. (1969), 'On the interpretation of discriminant analysis', *J. Marketing Res.*, **6**, 156–163.

Muxworthy, D. T. (1972), Review of AID III, *British Sociological Association Maths, Statistics and Computing Applications Group Newsletter*, **9**.

Orr, L. (1972), 'The dependence of transition properties in the educational system on observed social factors and characteristics', *J.R. Statist. Soc. (A)*, **135**, 74–95.

Rao, C. R. (1965), *Linear Statistical Inference and its Applications*, London: Wiley.

Scott, A. and Knott, M. (1974), 'A cluster analysis method for grouping means in the analysis of variance', *Biometrics*, **30**, 507–512.

Silverman, R. (1971), *A Comparison of Environmental Housing Standards and House Prices in a High Density Residential Area*, M.Sc. Thesis, University of Edinburgh.

Sokal, R. R. (1961), 'Distance as a measure of taxonomic similarity', *Syst. Zool.*, **10**, 70–79.

Sonquist, J. A. (1970), *Multivariate Model Building*, Ann Arbor: Institute for Social Research, University of Michigan.

Sonquist, J. A., Baker, E. L., and Morgan, J. N. (1971), *Searching for Structure*, Ann Arbor: Institute for Social Research, University of Michigan.

Sonquist, J. A., and Morgan, J. N. (1964), *The Detection of Interaction Effects*, Institute for Social Research, University of Michigan.

Sonquist, J. A. (1969), 'Finding variables that work', *Public Opinion. Quart.*, **23**, 83–95.

*Snowbarger, M., and Dunkelberg, B. (1970), 'A cross-sectional study of used car prices, 1962–4', *J. Marketing Res.*, **7**, 493–497.

Thompson, V. R. (1972), 'Sequential dichotomization; two techniques', *The Statistician*, **21**, 181–194.

Yates, F. (1934), 'The analysis of variance with unequal numbers in the different classes', *J. Amer. Statist. Ass.*, **29**, 51–66.

Author Index

The page numbers preceded by **I**: refer to Volume I; those preceded by **II**: refer to Volume II.

Subject Index

The page numbers preceded by I: refer to Volume I: those preceded by II refer to Volume II.